THE FUTURE OF THE BORNEAN ORANGUTAN
Impacts of change in land cover and climate

Editors
Serge Wich, Matthew Struebig, Johannes Refisch,
Andreas Wilting, Stephanie Kramer-Schadt and Erik Meijaard

UNEP

United Nations Environment Programme 2015

ACKNOWLEDGEMENTS

Editors

Serge Wich — Liverpool John Moores University, Liverpool (LJMU), UK, and University of Amsterdam, Amsterdam, The Netherlands

Matthew Struebig — Durrell Institute of Conservation & Ecology (DICE), University of Kent, Canterbury, UK

Johannes Refisch — Great Apes Survival Partnership (GRASP), UNEP, Nairobi, Kenya

Andreas Wilting — Leibniz Institute for Zoo and Wildlife Research (IZW), Berlin, Germany

Stephanie Kramer-Schadt — Leibniz Institute for Zoo and Wildlife Research (IZW), Berlin, Germany

Erik Meijaard — People and Nature Consulting, Jakarta, Indonesia

List of additional authors

Manuela Fischer — Leibniz Institute for Zoo and Wildlife Research (IZW), Berlin, Germany

David Gaveau — Center for International Forestry Research (CIFOR), Bogor, Indonesia

Catherine Gonner — Durrell Institute of Conservation & Ecology (DICE), University of Kent, Canterbury, UK

Rachel Sykes — Durrell Institute of Conservation & Ecology (DICE), University of Kent, Canterbury, UK

List of reviewers

Igone Arrillage — UNEP REDD, Nairobi, Kenya

Doug Cress — GRASP, Nairobi, Kenya

Laura Darby — GRASP, Nairobi, Kenya

Thomas Enters — UNEP RoAP Office, Bangkok, Thailand

Milja Fenger — Faculty of Science, University of Copenhagen, Denmark

Lera Miles — UNEP WCMC, Cambridge, UK

William Olupot — Bushmeat-Free Eastern Africa Network (BEAN), Kampala, Uganda

Tania Salvaterra — UNEP WCMC, Cambridge, UK

Makiko Yashiro — UNEP RoAP Office, Bangkok, Thailand

Max Zieren — UNEP RoAP Office, Bangkok, Thailand

CONTENTS

Rehabilitation at Pulau Kaja, Sungai Rungan, Central Kalimantan
Photo: Perry van Duijnhoven

FOREWORD

Achim Steiner, United Nations Under-Secretary-General and Executive Director of the United Nations Environment Programme. **Photo** UNEP

Over the past century, orangutan populations in Southeast Asia have seen a very steep decline, driven to the brink of extinction by a host of man-made threats. Deforestation, illegal logging, the expansion of agro-industrial plantations and hunting – these forces combined to isolate orangutans into precarious pockets of forest on the islands of Borneo and Sumatra.

Now, a new threat has emerged: climate change. This report from the Great Apes Survival Partnership (GRASP) and the Liverpool John Moores University assesses the impacts of land cover change and climate change on Borneo's endangered orangutans. The report also examines the major driver of deforestation – the expansion of oil palm – and analyses how various land-use scenarios might impact the region through different climate change projections. The report concludes, sadly, that a combined model of climate change and landuse change could result in a further three quarter loss of orangutan habitat from the present day.

The United Nations Environment Programme (UNEP) actively seeks solutions to reverse this trend. The 2014 UNEP report, *Building the Natural Capital*, indicates that if systematically pursued, REDD+ (Reducing Emissions from Deforestation and Degradation) could both address climate change and preserve some of the world's tropical forests, while also protecting biodiversity. Key findings show that REDD+ can also correct markets, policy, and institutional failures that undervalue the more serious climate change mitigation services provided by ecosystems.

GRASP also addressed this opportunity in the 2011 report, *Orangutans and the Economics of Sustainable Forest Management in Sumatra*, which called for "vigorous efforts" to reverse the impact of climate change, and stated, "the development of a Green Economy can lead to a win-win situation where orangutan habitat is conserved, ecosystem services maintained and economic growth continued."

UNEP supports global initiatives such as the Economics of Ecosystems and Biodiversity, the Intergovernmental Panel on Biodiversity and Ecosystem Services, and the Natural Capital Declaration, each of which provides new tools to measure the value of ecosystem services to our well-being. Now, it is time to utilize these approaches and divert from an unsustainable pathway to development. *The Future of the Bornean Orangutan: Impacts of Change in Land Cover and Climate* makes it clear that a future without sustainable development will be a future with a different climate and, eventually, without orangutans, one of our closest relatives.

Mother with infant, Tuanan, Central Kalimantan.

SUMMARY

Traditionally the main threats to biodiversity have come from changes to land cover as well as hunting. Recently another threat has become apparent: climate change. An increasing number of studies suggest that climate change is having negative impacts on biodiversity and that these impacts will increase in the future. It is therefore important to assess whether the impact of climate change will add to the existing threat of land-cover change, particularly in regions where habitat loss is high and a large proportion of the world's biodiversity resides; regions such as the island of Borneo. In this report the impacts of land cover and climate change are assessed for one of Borneo's most iconic species, the Bornean orangutan.

The report assesses the uncertainty in climate model projections and how this affects predictions of climate change on Borneo for the years 2020, 2050 and 2080. Projections from two climate models (CSIRO-Mk2 and HADCM3) are examined under two emission scenarios (A2 and B2), which reflect 'worst-case' and 'best-case' trajectories of global development.

This best-case scenario represents a future oriented towards environmental protection and social equity: slower population growth, less land-cover change, and more diverse technological developments. The clearest climate pattern for Borneo over time is a projected rise in average annual temperature, which is particularly evident along the coastal lowlands where most of the island's urban populations reside. According to the models assessed, island-wide mean temperatures could rise by 2-3 °C above current levels by 2080.

Climate models also indicate a greater variability in future temperatures, as indicated by an increase in annual temperature range and seasonality over time. These changes are particularly pronounced in the HADCM3 climate-model outputs. Over the course of this century Borneo is likely to experience elevated rainfall, but the magnitude of this change varies substantially amongst the two climate-model projections. Under the CSIRO model projections, total annual rainfall over the Borneo lowlands could increase by up to 224-520 mm by 2050 and 513-754 mm by 2080, representing an increase of more than 15% of present levels in some parts of the island, including the large cities of Banjarmasin, Kota Kinabalu, Kuching, Palangkaraya and Samarinda. However, the Hadley model predictions for rainfall are generally less severe, pointing to an increase of, at most, 248-305 mm by the end of the century.

This report also explores the potential impact of land-cover and climate change on the distribution of suitable orangutan habitat by using bioclimatic and habitat suitability modelling. Models using recent climate and land-cover conditions for 2010 indicate that up to 260 000 km^2 of land, including intact but also degraded and fragmented forest, could be considered potential orangutan habitat within the known species range. This figure represents an 18% reduction of the former core range due to land cover changes, according to the same criteria but using models repeated to land-cover conditions prior to the 1950s.

Climate parameters which had the greatest influence on the model are: 1) annual temperature range, 2) precipitation in the driest month and quarter, and 3) mean diurnal temperature range. According to projections of these models to future climate conditions 49-69% of this habitat would become unsuitable by the 2080s. Much of this retraction is projected to take place in southern part of the island in Central Kalimantan, though there is some variation in the extent of this loss between the climate forecasts. Conversely, there is little difference between the outcomes of the two emission scenarios examined.

Because rates of forest loss over Borneo are high it is also important to assess what the impact of future deforestation on orangutan habitat could be. Repeating the suitability assessment using land cover derived from a predictive model of deforestation an estimated 251 000 km^2 of habitat is predicted to remain suitable by 2020, leading to 219 000 km^2 by 2080 and representing an overall reduction of suitable habitat of around 15% by the end of the century. Much of this forest loss is predicted to take place in the lowlands of the island, principally in the areas of Sarawak and Kalimantan, as it has over the last ten years.

Climate change and land cover are thus predicted to adversely affect similar regions of Borneo, at least for some parts of the island. The combined impact of land cover and climate change might therefore be more pronounced than either factor alone. Suitability models incorporating both projected changes to climate and land cover indicate that up to 83 000 km^2 would remain suitable by 2080, representing 68-81% loss from the present day. The loss of orangutan habitat due to the combined effects of land cover and climate change could be more than double the amount lost historically and also considering land-cover projections alone.

The report also includes an appraisal of land suitable for oil palm and asks how the extent of this area might change under climate change projections, and whether the change might have implications for orangutan conservation. The main finding is that with changing rainfall patterns predicted for this century there will likely be a reduction in the productivity of large expanses of land for oil-palm cultivation over the island. These low productivity lands primarily encompass the north and south-central parts of the island, where much of the recent expansion of this crop has taken place.

Large areas, however, including regions in West Kalimantan and Sabah where the crop is already well-established, will remain suitable. Overlaying these areas with those identified as suitable for orangutans gives us insight into regions where land-use conflicts are likely to occur for this species over the course of the century. These regions include some of the strongholds for orangutans across the various environmental change projections – in central Sabah, East Kalimantan's Sangkulirang-Mangkalihat peninsula, and the Kapuas lakes and foothills of the Schwaner mountains in West Kalimantan.

RECOMMENDATIONS

1   Identify all remaining orangutan populations and undertake an assessment of their viability, based on local threat levels from killing and incompatible land use, present and predicted habitat quality, and population dynamics (dispersal, inbreeding, etc.). Identify all orangutan populations that the governments officially agree to maintain.

2   Conduct a triage process in which the 'must save' and 'can save' populations are identified and agreed by governments, both inside and outside protected areas, as well as the management measures needed to maintain them.

3   Based on the above, propose and designate new protected areas or other areas under permanent forest cover that are large and safe enough to contain viable orangutan populations.

4   Where possible, connect these permanently forested areas through uninterrupted forested corridors, for example in permanent natural forest timber concessions, that allow orangutans and other wildlife to move through the landscape in reaction to changing climatic and ecological conditions.

5   Reconcile these land-use plans with other spatial plans (for development, infrastructure, agriculture, etc.), and endorse these planned land uses in high level government regulations that allocate special strategic status to these orangutan populations, and which are strong enough not to be overruled by other national-level or local level regulations.

6   Effectively enforce laws regarding the killing of orangutans and implement public campaigns and other communication efforts that make the public aware of the illegality of killing.

7   Seek innovative ways to augment protected areas to conserve remaining orangutan forests, which could include, for example, leverage of carbon mitigation investments via the UN-REDD+ (Reducing Emissions from Deforestation and Forest Degradation) mechanism, or other payments for forest environmental services (e.g., erosion control, flood buffering, local climate regulation). Such systems might not only reduce threats to biodiversity from land-cover changes but also contribute to local community welfare and safety and to international efforts to reduce climate change.

8   Drained coastal peatlands in Borneo are predicted to decompose and flood leaving behind unproductive brackish swamps. Peatlands are key orangutan habitats that should be left forested and undrained to avoid major negative biodiversity and socio-economic impacts.

Figure 1 **Location of Borneo in Southeast Asia and the major administrative boundaries and cities.**
The island is governed by three countries: Brunei Darussalam, Indonesia (five provinces) and Malaysia (two states).

BORNEO

INTRODUCTION

The future of most wildlife is now inextricably linked to the political, economic and social choices of one species: humans. Species and ecosystems are increasingly being affected negatively by human activities which have resulted in widespread pollution, overharvesting, habitat destruction and fragmentation. Some researchers predict that the human race could be responsible for a "sixth mass extinction", with a large portion of species going extinct during this century (Barnosky *et al.* 2011). Many of the projected extinctions of terrestrial species are likely to take place in the humid tropics. Tropical rainforests, which cover less than 7% of all land, not only host much of the world's biodiversity but are also experiencing large-scale loss (Hansen *et al.* 2013; Myers *et al.* 2000). The presence of many specialist and endemic species in these regions creates great risk of extinction due to deforestation. For example, 59.6% of the 29 375 vascular plant species in Indonesia do not occur anywhere else on the planet and local disappearance would

54 000 individuals left in the wild (Wich *et al.* 2008, Figure 2). The population has seen a sharp decline over the past few decades (Rijksen & Meijaard 1999) and continues to suffer from human disturbances such as expanding agriculture, hunting, logging, and the pet trade (Stiles *et al.* 2013; Wich *et al.* 2012; Meijaard *et al.* 2011; Wich *et al.* 2008; Nellemann *et al.* 2007). While the most immediate threat to orangutans and other Bornean wildlife is the continuing loss of forest habitat and hunting, it is now recognised that climate change will likely play an important detrimental role in their long-term futures as well (Struebig *et al.* 2015; Gregory *et al.* 2012). These threats combined could seriously hinder the considerable attempts by conservationists to protect species from extinction. The Indonesian government has expressed its wish to stabilize all wild orangutan populations in Indonesia by 2017 (Departemen Kehutanan 2007) and similar plans have been proposed by the Sabah Wildlife Department (Sabah Wildlife Department 2011).

Borneo   *After Greenland and New Guinea, Borneo is the third largest island in the world, with a land area of 743 330 km², i.e. larger than the state of Texas or the entire Iberian Peninsula in Europe. Nearly 20 million people live in the Bruneian, Indonesian, and Malaysian parts of the island (Figure 1), resulting in an average population density of about 25 people/km², which is comparable to Brazil or Sweden and much lower than the U.S.A. Most of Borneo's population lives in coastal cities, although there are many small towns and villages further inland.*

---

mean global extinction (Sodhi *et al.* 2004). Of all areas in the humid tropics, Southeast Asia experiences the highest relative rate of deforestation (Achard *et al.* 2002) and a recent analysis showed that globally, Indonesia exhibited the largest increase in forest loss from 2000-2012 (Hansen *et al.* 2013). If deforestation in the region continues, around 75% of the original forest cover could be lost by 2030 (Brook *et al.* 2006). The Southeast Asian island of Borneo has an estimated 14 423 plant and 1 640 vertebrate species, of which 4 508 are endemic and 534 threatened with extinction (IUCN 2013). Borneo is home to a number of global conservation icons such as the critically endangered Sumatran rhinoceros (*Dicerorhinus sumatrensis*) and the endangered Borneo pygmy elephant (*Elephas maximus borneensis*).

Even better known is the endemic and endangered Bornean orangutan (*Pongo pygmaeus*). This Asian great ape primarily dwells in lowland forests at low densities and recent estimates suggest that there are approximately

Success in ensuring the survival of such iconic species such as the orangutans has the potential to galvanize conservation efforts within Indonesia and beyond.

Land-cover changes

Borneo's forests are among the most diverse in the world. The vegetation on the island varies in species composition and structure because of factors such as altitude, soil characteristics (*e.g.* forest on peat or near beaches), and climate variation (MacKinnon *et al.* 1996). In 1973, around the time when large-scale commercial logging began, an estimated 76.4% of Borneo's area was forested, but by 2010 about 30% of these forests had been cleared and a further 31.6% had been 'logged', *i.e.* economically valuable trees had been extracted (Gaveau *et al.* 2014, Figure 3). Initially much of this large-scale deforestation and degradation was caused by the extraction of the commercially important dipterocarp tree species for durable hardwood, but in recent years the conversion of degraded forest to monoculture plantations has played a greater role. By 2012, about 10% of Borneo's land area had been converted to industrial plantations of, for example, oil palm, pulp and paper, and timber (GAVEAU *et al.* 2014). Such land-cover changes have had obvious impacts on wildlife. The orangutan might, in some areas, survive at much-reduced densities in plantation areas, especially if patches of natural forest remain (ANCRENAZ *et al.* 2014), but their densities are far lower than those found in carefully managed timber concessions and unlogged forests (ANCRENAZ *et al.* 2010; HUSSON *et al.* 2009). Furthermore, their survival in such human-dominated areas may be further compromised; the presence of roads and villages can facilitate hunting, poaching for the pet trade and killing of individuals because of agricultural conflict. In Kalimantan alone between 2 000 and 3 000 orangutans may be killed each year in this way (MEIJAARD *et al.* 2011).

Climate change

It is now widely accepted that this century will see a rapid shift in the global climate (IPCC 2013). The changes will likely include a rise of average global temperatures by between 1.5 and 2 °C by 2100, alterations in precipitation patterns and an increase in the frequency of extreme weather events (IPCC 2013). Accompanying these changes will be the continued loss of ice in glaciers and polar regions and a rise in sea levels of up to 0.98 m by the end of the century (IPCC 2013). The effects of such wide-ranging environmental changes on wildlife could be devastating. One study predicted that in the coming decades climate change may become a greater threat to biodiversity than habitat loss (LEADLEY 2010).

There are a great number of ways that future climate change may affect the survival of species. Investigations into these effects have only recently begun in earnest; the actual existing and potential impacts of global climate change on species and ecosystems are now being extensively explored by ecologists in order to gain an understanding of the biological mechanisms involved (BELLARD *et al.* 2012). At the level of the ecosystem, the greatest impacts may be expected in regions closer to the poles, which is where temperature and rainfall patterns will be most altered. In the tropics, shifts in wildfire patterns will also play a role, as well as the effect of CO_2 fertilization, the process by which increased levels of atmospheric CO_2 helps plants build up biomass quickly (THUILLER 2007). However, in terms of individual species, changing local weather patterns in the tropics will determine in part where, and in what numbers, species can survive. Climatic factors, in particular mean annual temperature and wet season precipitation, affect orangutan abundance and distribution in Sabah (GREGORY *et al.* 2012), so changes in the local climates in Borneo are expected to have a direct effect on orangutan populations.

Modelling changes into the future

Without predicting future responses to the changes in climate that may occur in the coming century, designing an effective conservation strategy for the Bornean orangutan would be difficult. Through modelling change into the future, policymakers can be warned about dangers to the orangutan population, and conservation areas can be designed rationally with the future in mind (MEIJAARD *et al.* 2012). For example, if there are plans for a new conservation area in a region where orangutans fare well now but where climate models show that they will not do well in the future, these plans should ideally be reconsidered.

The future distribution of orangutans

This report uses both climate models and Borneo-specific land-cover change models to project scenarios of change from now until 2080 in the extent of suitable habitat for the orangutan over the entire island (STRUEBIG *et al.* 2015). In this way it is possible to assess simultaneously how the survival of the Bornean orangutan might be affected by these two overarching threats of environmental change. This will help us create a more realistic image of the prospects for the orangutan species range and thus the future needs for its conservation. The urgency for implementation of effective orangutan conservation measures cannot be overemphasized: with a shrinking habitat and thousands of individuals being killed each year there is a realistic possibility that the Bornean orangutan may, within decades, be confined to a very limited number of small and isolated populations.

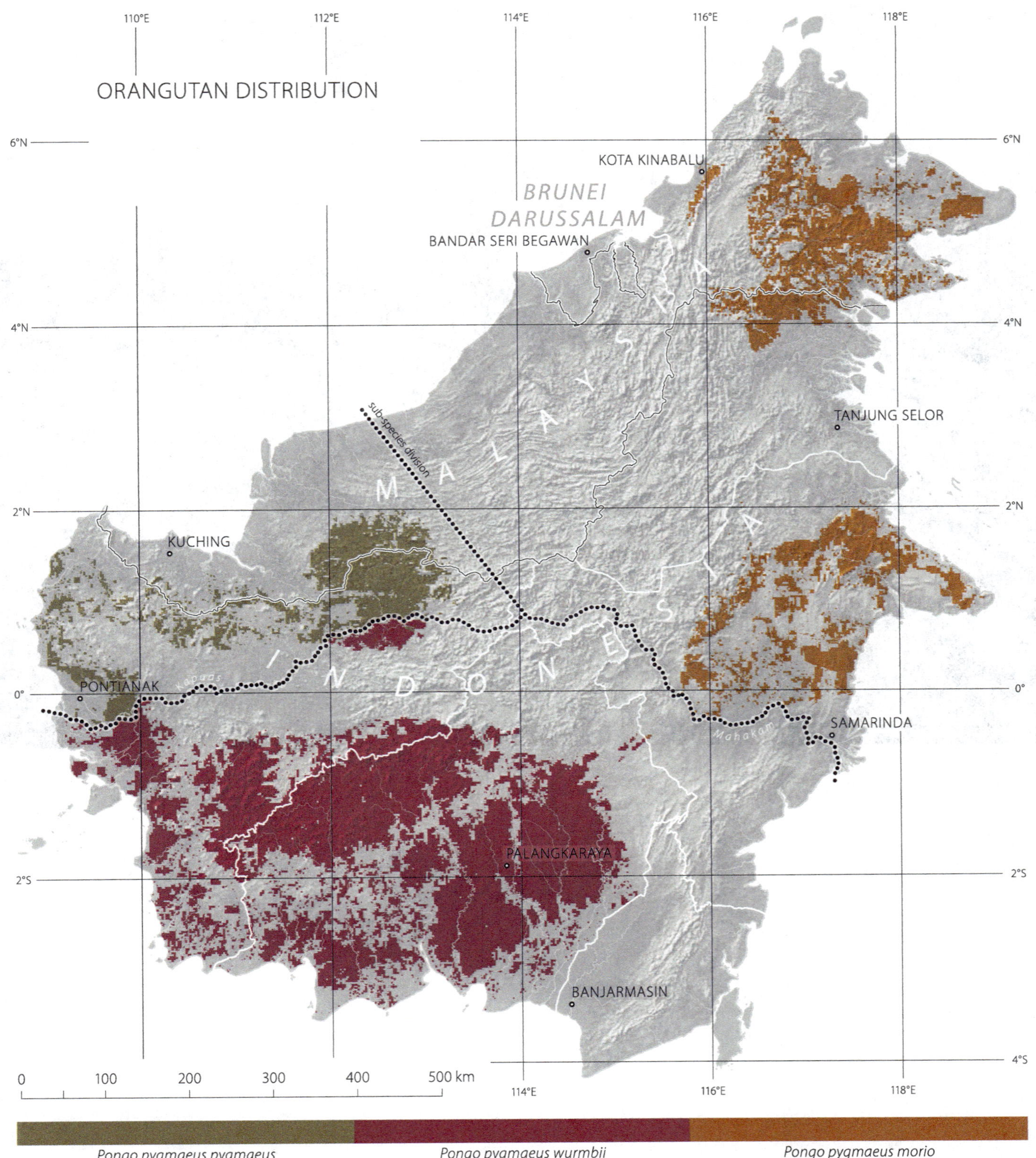

Figure 2 **The Bornean orangutan is subdivided into three subspecies.**
Predicted core distribution of the three Bornean orangutan subspecies, *P. p. pygmaeus*, *P. p. wurmbii*, and *P. p. morio* based on Wich *et al.* (2012).

Drainage canals and destroyed peat swamp forest, legacies from Suharto's Mega Rice Project in Central Kalimantan. **Photo** Perry van Duijnhoven

Log rafts travelling down the Barito river, near Buntok, Central Kalimantan. **Photo** Perry van Duijnhoven

Structure of this report

There are a number of important issues that need to be considered in order to assess and understand the potential impacts of climate change on the Bornean orangutan. There is first a need to recognize the uncertainty of the actual climatic changes that will occur on the island. In this document, different climate models as well as different scenarios of global greenhouse gas (GHG) emissions are examined. The models examined are the CSIRO-Mk2 model from the Commonwealth Scientific and Industrial Research Organisation of Australia and the HADCM3 model from the Hadley Centre for Climate Prediction and Research in the UK. For each of these models the scenarios examined are a 'worst-case' and a 'best-case' scenario with the latter representing a future oriented towards environmental protection and social equity: slower human population growth, less land-cover change, and more diverse technological developments, and the worst-case scenario being a projection of 'business as usual'. Potential range shifts or contractions that might occur because of changes to bioclimatic suitability for the orangutan under the resulting future climate scenarios are studied next, followed by the impact of land-cover change on the extent of suitable habitat for orangutans. After this the scenarios for climate change and land-cover change are combined to assess their impact on the extent of suitable habitat for orangutans. This is then followed by analyses that assess the potential effects of climatic changes on the suitability of land for palm-oil production. The expansion of the oil-palm sector is a leading threat for the conservation of orangutan, and so any evaluation of areas likely to support this species in the future should also consider possible land-use conflict with this economically important crop.

15

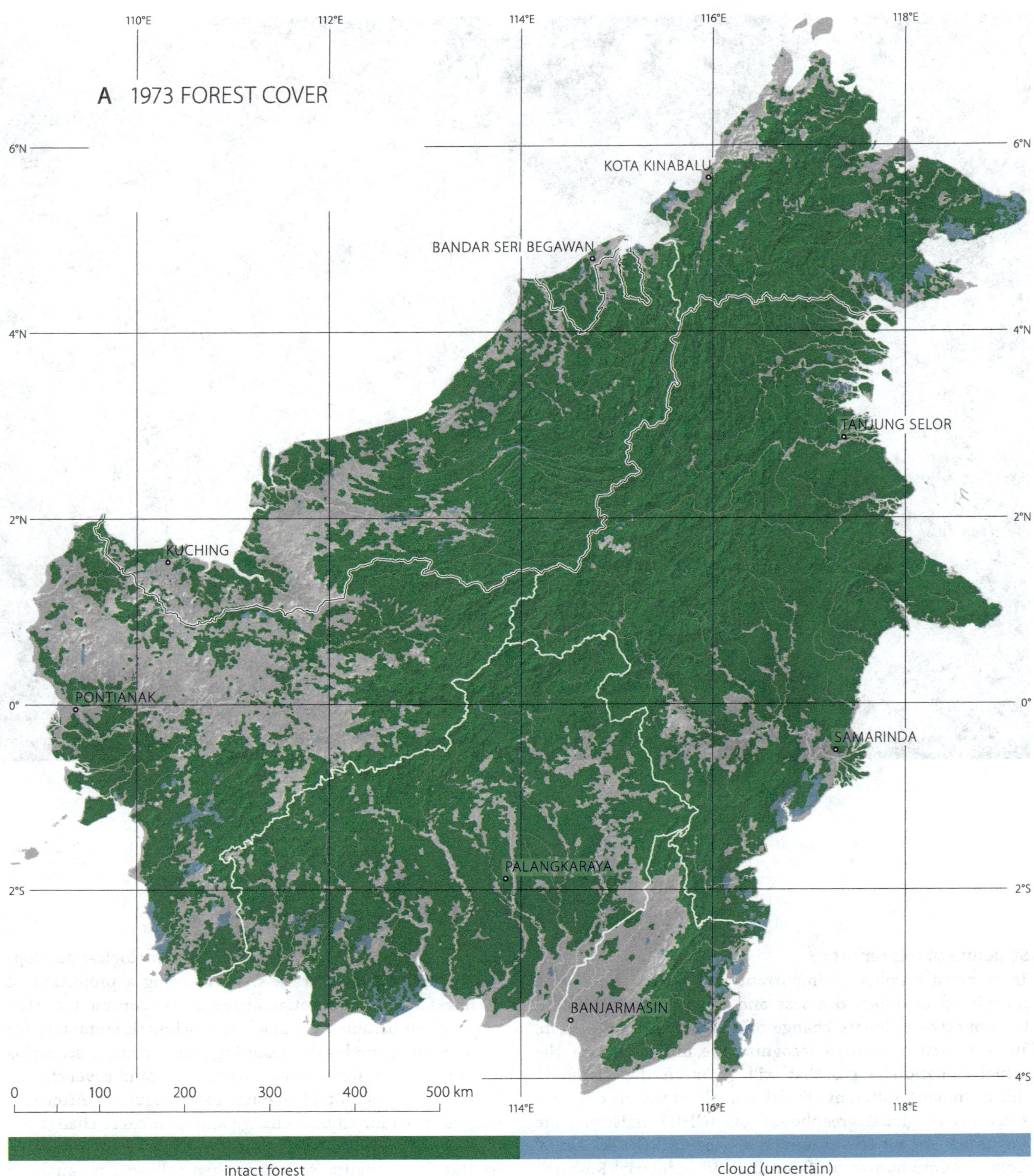
A 1973 FOREST COVER
110°E
112°E
114°E
116°E
118°E
6°N
KOTA KINABALU
BANDAR SERI BEGAWAN
4°N
TANJUNG SELOR
2°N
KUCHING
0°
PONTIANAK
SAMARINDA
PALANGKARAYA
2°S
BANJARMASIN
4°S
114°E
116°E
118°E
0
100
200
300
400
500 km
intact forest
cloud (uncertain)

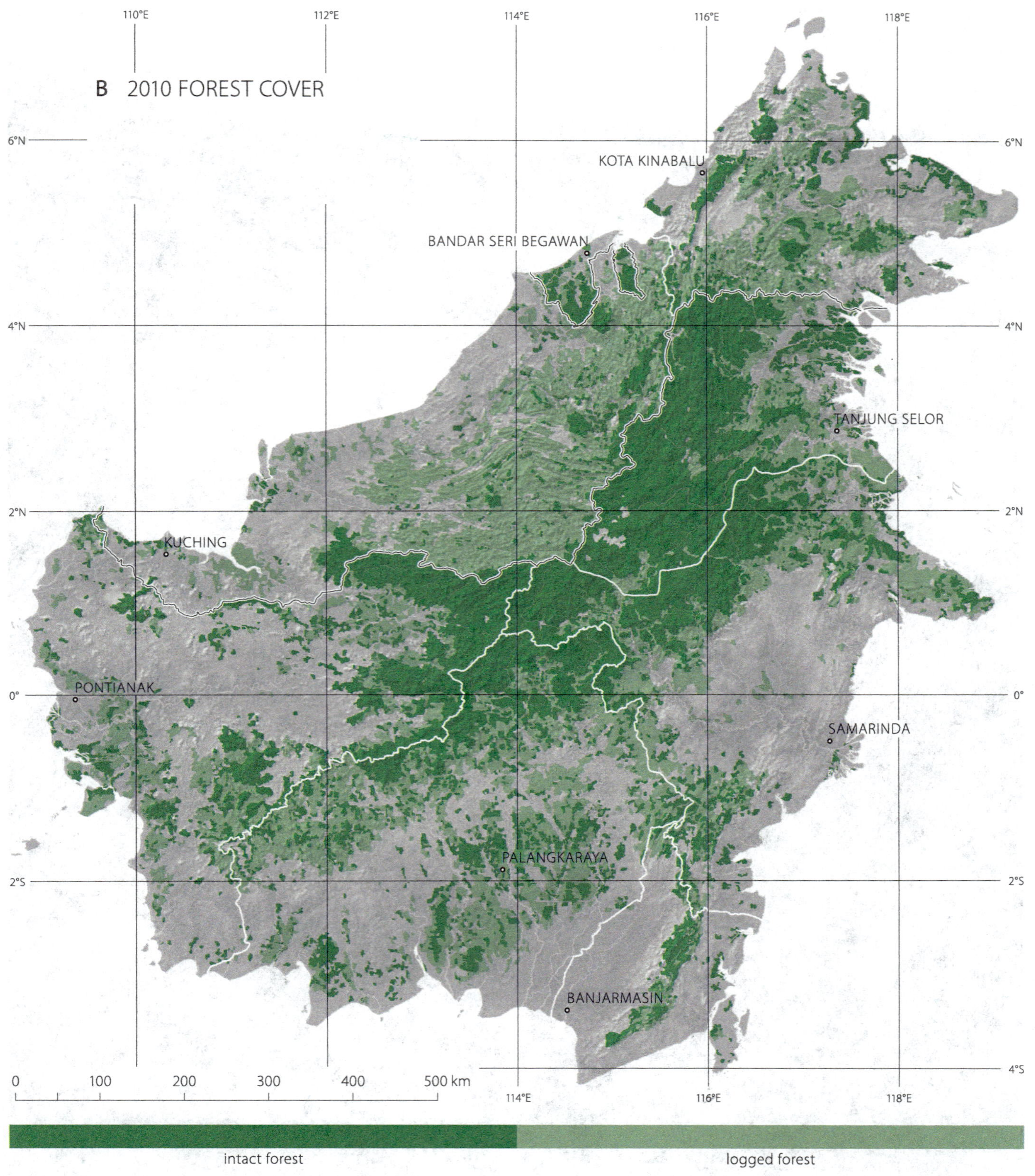

Figure 3 **Forest disturbance and loss due to logging and plantation development on Borneo.**
Maps show forest cover in 1973 (A) compared to 2010 (B) (Gaveau *et al.* 2014).

Forest canopy, Brunei.

CLIMATE CHANGE ON BORNEO

The planet's climate is affected by changes in atmospheric abundance of greenhouse gases, land surface properties, and solar radiation. Climate has changed and will continue to change over time due to natural variability, but importantly, can also be attributed to human activities. The burning of fossil fuels and changes in land cover since 1750 have led to

Change (IPCC), a scientific body set up by the United Nations in 1988. The IPCC is comprised of thousands of leading scientists who act voluntarily to review current scientific and socio-economic climate data, and advise the governments of more than 120 countries. As such the IPCC is the internationally accepted authority on climate change, and each report it

---

***Determining climate change** Climatologists use temperature measurements taken over land and water across the globe to investigate how temperature has changed over time and how climate varies regionally. For example, observations of surface air temperatures for the Indo-Malayan region during the last century indicate that a temperature increase of around 0.5 °C has occurred, a magnitude that is in broad agreement with global estimates of an increase of 0.7 °C ± 0.2 (BAPPENAS 2010). As few weather stations exist on Borneo compared to other parts of the world, much of the information on the climate on the island is actually based on interpolation using data from the nearby Malay Peninsula, Philippines and Java (Figure 4). Climate scientists have developed models to project changes in climate into the future, and evaluate the influence that climate might have on the global environment, including on biodiversity. Such climate projections are undertaken using global General Circulation Models (GCMs), which simulate climate processes under various physical assumptions, and the responses of the climate system to elevated greenhouse gas concentrations. These climate models can be thought of as highly sophisticated extensions of models used to predict weather, though they are still in their infancy, with the first model having been established in 1965. In preparation for the IPCC Fifth Assessment report, scientists used up to 45 different GCMs from 24 modelling centres across the world, all operating under different climatological assumptions and at different time intervals. This way they can surpass problems with uncertainty associated with any given model. So, for any given time interval (or time-slice), projected climate data are available from a number of different GCMs, and each model simulation might be expected to yield different results because of different underlying model assumptions. Traditionally, biological research on the potential impacts of climate change has been limited to using the outputs from a single GCM. This method is of concern because subtle differences in predicted local climates across models might lead to differing conclusions in, for example, the projected suitability of a habitat for a specific species. Only recently, with advances in computing, has it become possible to examine the outcomes of using data from multiple climate models.*

---

elevated concentrations of anthropogenic greenhouse gases (namely carbon dioxide, methane and nitrous oxide) in the atmosphere. The concentration of carbon dioxide alone has increased by more than 70% compared to pre-industrial levels (IPCC 2013). Assessing the risks associated with climate change and its potential socio-economic and environmental impacts is the role of the Intergovernmental Panel on Climate

produces is discussed and approved by leading climate scientists and participating governments. The Fourth Assessment Report (AR4) was published in 2007, and concluded that the warming of the climate system is "unequivocal", and that it is "very likely" due to an observed increase in anthropogenic greenhouse gas concentrations. The Fifth Assessment Report (AR5) was near completion at the time of writing this report.

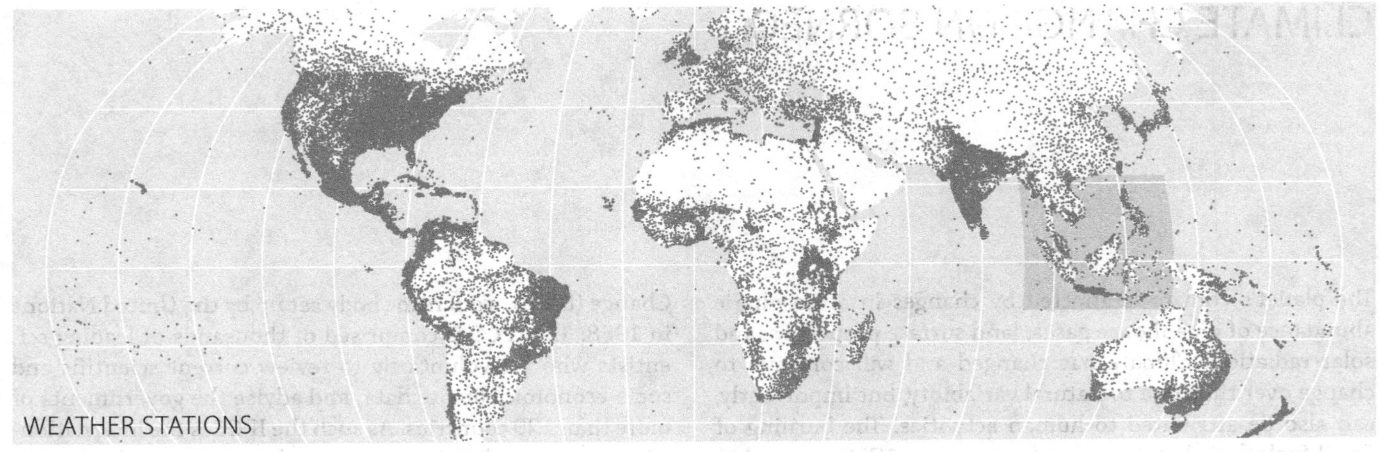

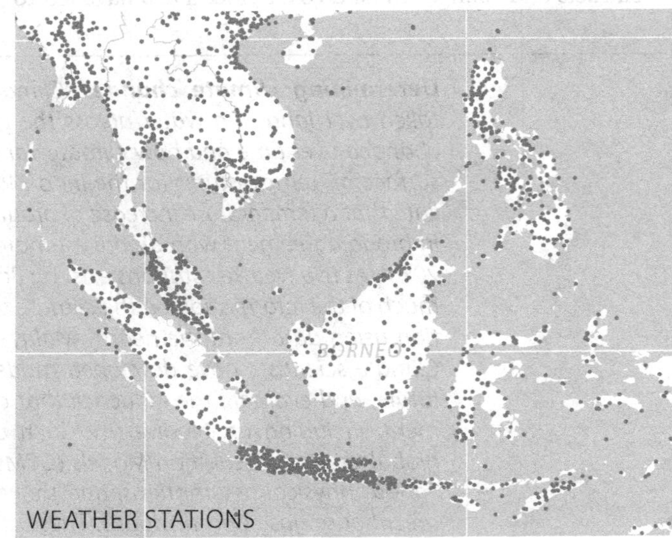

Figure 4 **Locations of weather stations used to map rainfall and temperature data between 1950 and 2000.**
Measurements collected from the world's weather stations are used to interpolate, or estimate, data at under-sampled areas, in order to map climate data over the global surface. In recent years this has led to the creation of a publically available global spatial dataset compiled from monthly averages of 47 554 weather station measurements for the 1950-2000 time period (HIJMANS *et al.* 2005).

Elevated temperatures have knock on effects on precipitation patterns and sea levels. This chapter presents information on Borneo's climate in the past and the future.

Borneo's current and past climate

Borneo straddles the equator and mostly has a tropical, moist climate. Temperatures are relatively constant throughout the year, varying between 25 °C and 35 °C in most of the lowlands. Near-zero temperatures are only experienced in the highest mountain areas, such as the 4 000 m tall Mount Kinabalu in Sabah. Rainfall patterns are influenced by monsoonal winds, with south-easterly winds prevailing between May and October and north-westerly ones between November and April. This weather pattern generally results in a drier season between July and October, which is most pronounced towards the south-east of the island (Figure 5). Still, Borneo has very few months with less than 200 mm of rainfall. Most of the hilly inland areas receive between 2 000 and 4 000 mm of rain each year (MACKINNON *et al.* 1996).

Climate variability in Borneo, both in time and space, has been well-recognized. Despite having an equatorial climate there are subtle differences in conditions across the island. A long-term data series of average annualized monthly rainfall from Borneo weather stations (Figure 5) indicates significant fluctuation between different years, with some years totalling as much as 4 500 mm of rain averaged over 20 different rain stations on the island, and others only 2 500 mm. Although the overall rainfall trend between 1876 and 2011 appears to be relatively stable, at a finer scale there are indications of overall rainfall increasing since the late 1990s. Because of the climatic variability across the island, there appear to be pronounced trends in rainfall change over time in some areas. For example, all three rainfall stations in southwestern Borneo (Singkawang, Pontianak, Ketapang) show declining rainfall trends between 1876 and 2011, with especially 'wet-season' December rainfalls apparently declining. Over much longer time-scales Borneo has experienced a range of different climatic conditions; it has not always been as wet as it is now. Over 25 million years there have been alternations between ever-wet and more seasonal climates, sometimes

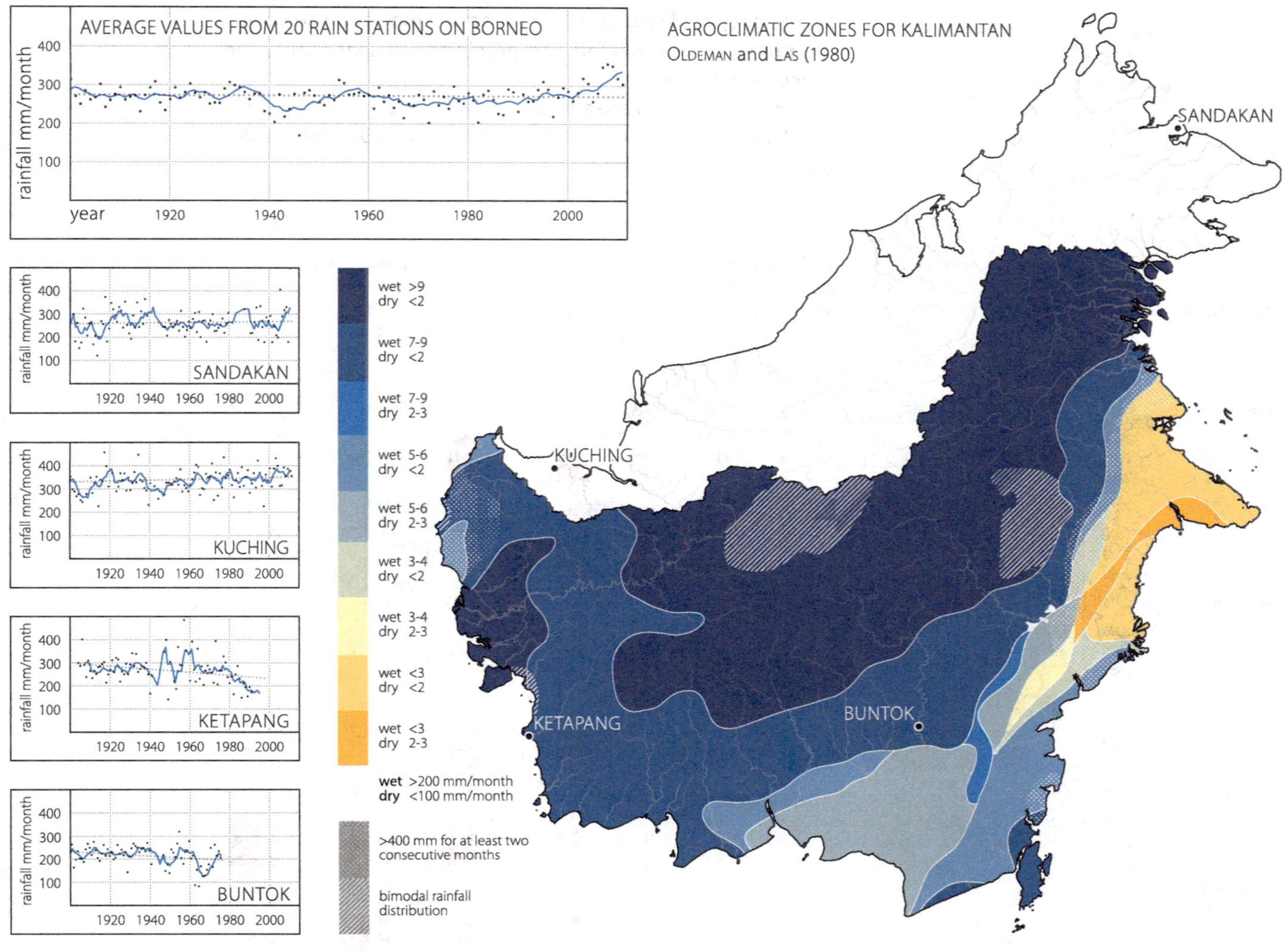

resulting in quite dry conditions. For example, in recent ice-ages, parts of eastern Borneo were covered in savanna-like vegetation (FLENLEY 1998; CARATINI & TISSOT 1988). The interior of Borneo, however, seems to have maintained rainforest conditions throughout these times (ANSHARI *et al.* 2001).

Modelling the future climate

The remaining part of this chapter presents informa-tion on projected temperature and rainfall patterns in Borneo for three future time periods – the 2020s, 2050s and 2080s (STRUEBIG *et al.* 2015). These projections aim to take into account variation in different climate models and future greenhouse gas (GHG) emissions. The use of General Circulation Models (GCMs) is essential in order to predict local climate patterns. The projections produced from two different GCMs were used to represent the range in predictions from four GCMs available at high resolu-tion for the time periods under study. Both are commonly applied in studies of future climate for Southeast Asia.

Figure 5   Annualized monthly rainfall from rain stations on Borneo, between 1900 and 2011.

The top graph indicates the average values from 20 rain stations over Borneo, and other graphs illustrate local variation in rainfall trends. For all graphs the dotted grey lines indicate the overall linear trend, while the blue lines indicates a 5-year moving average. Data compiled by EM from Southeast Asian Climate Assessment & Dataset (SACA&D). Note, data outliers, such as the low record in 1946 for the main graph are influenced by missing data. These data are preliminary, however, and some caution in their interpretation is needed because data quality has not yet been assessed. The central map shows the agroclimatic zones for Kalimantan (OLDEMAN & LAS 1980).

Table 1 **Projected change in annual mean temperature, temperature range, annual rainfall and precipitation seasonality for Borneo's main cities.**
Climate values, magnitude and direction of climate change is indicated for the two emission scenarios under the CSIRO global circulation model.

city	year	annual mean temperature worst-case	Δ	best-case	Δ	temperature range worst-case	Δ	best-case	Δ	annual rainfall worst-case	Δ	best-case	Δ	precipitation seasonality worst-case	Δ	best-case	Δ
Bandar Seri Begawan	2010	27.6 °C				7.4 °C				3070 mm				27 %			
	2020	28.3		28.3		7.4		7.6		3174		3231		27		28	
	2050	29.0	+2.5	28.9	+1.9	7.4	+0.2	7.5	+0.3	3204	+381	3335	+300	25	-4	25	-2
	2080	30.1		29.5		7.6		7.7		3451		3370		23		25	
Banjarmasin	2010	26.7 °C				10.6 °C				2414 mm				43 %			
	2020	27.5		27.5		10.9		11.0		2611		2597		44		43	
	2050	28.2	+2.7	28.2	+2.1	11.0	+0.5	11.1	+0.3	2613	+435	2852	+365	46	+4	44	-3
	2080	29.4		28.8		11.1		10.9		2849		2779		47		40	
Bintulu	2010	26.5 °C				8.7 °C				3791 mm				20 %			
	2020	27.2		27.2		8.9		9.0		4002		4074		19		23	
	2050	27.9	+2.5	27.8	+1.9	8.9	+0.3	9.0	+0.2	4025	+583	4223	+427	23	0	18	-4
	2080	29.0		28.4		9.0		8.9		4374		4218		20		16	
Ketapang	2010	26.9 °C				9.5 °C				3142 mm				35 %			
	2020	27.7		27.8		9.0		9.5		3208		3177		34		33	
	2050	28.5	+2.8	28.4	+2.1	9.5	+0.2	9.2	+0.6	3234	+261	3364	+251	36	0	34	-2
	2080	29.7		29.0		9.7		8.9		3403		3393		35		33	
Kota Kinabalu	2010	26.7 °C				8.9 °C				2731 mm				32 %			
	2020	27.6		27.7		8.9		8.7		2954		2983		36		35	
	2050	28.3	+2.7	28.3	+2.1	8.8	0	8.8	-0.1	2954	+453	3085	+399	31	+3	35	+3
	2080	29.4		28.8		8.9		8.8		3184		3130		35		35	
Kuching	2010	27.0 °C				9.9 °C				3715 mm				32 %			
	2020	27.7		27.7		10.0		10.1		4315		4245		39		39	
	2050	28.4	+2.4	28.3	+1.9	10.1	+0.2	10.2	0	4406	+1016	4480	+806	47	+12	43	+7
	2080	29.4		28.9		10.1		9.9		4731		4521		44		39	
Muara Teweh	2010	26.4 °C				8.3 °C				3141 mm				27 %			
	2020	27.2		27.2		8.5		8.6		3386		3406		31		39	
	2050	27.9	+2.6	27.8	+2.0	8.8	+0.6	8.7	+0.4	3414	+621	3711	+564	34	+10	43	+4
	2080	29.0		28.4		8.9		8.7		3762		3705		37		39	
Palangkaraya	2010	26.6 °C				9.7 °C				2605 mm				28 %			
	2020	27.4		27.4		9.5		9.8		2800		2787		32		28	
	2050	28.1	+2.7	28.0	+2.0	9.7	+0.2	9.7	-0.1	2796	+466	3055	+446	34	+9	32	+2
	2080	29.3		28.6		9.9		9.6		3071		3051		37		30	
Pontianak	2010	27.6 °C				9.7 °C				3161 mm				23 %			
	2020	28.3		28.4		9.5		9.8		3269		3213		20		21	
	2050	29.1	+2.6	29.0	+2.0	9.7	-0.1	9.7	-0.2	3277	+323	3359	+263	24	+1	21	-3
	2080	30.2		29.6		9.9		9.6		3484		3424		24		20	
Samarinda	2010	26.8 °C				7.7 °C				2102 mm				17 %			
	2020	27.6		27.6		8.1		7.8		2243		2253		21		16	
	2050	28.3	+2.6	28.3	+2.0	8.0	+0.7	7.9	+0.6	2246	+342	2426	+305	25	+12	22	+6
	2080	29.4		28.8		8.4		8.3		2444		2407		29		23	
Sandakan	2010	27.0 °C				9.5 °C				3060 mm				38 %			
	2020	27.7		27.8		9.5		9.5		3155		3222		38		38	
	2050	28.5	+2.5	28.4	+1.9	9.5	+0.2	9.7	+0.1	3208	+285	3269	+216	40	-3	37	-3
	2080	29.5		28.9		9.7		9.6		3345		3276		35		35	
Sintang	2010	26.9 °C				9.8 °C				3496 mm				15 %			
	2020	27.6		27.6		9.7		9.9		3686		3681		15		16	
	2050	28.3	+2.5	28.2	+1.9	9.7	0	9.7	-0.2	3677	+508	3888	+423	21	+8	18	+1
	2080	29.4		28.8		9.8		9.6		4004		3919		23		16	
Tanjung Selor	2010	26.9 °C				8.0 °C				2715 mm				13 %			
	2020	27.5		27.6		8.2		8.0		2817		2885		13		14	
	2050	28.2	+2.5	28.2	+1.9	8.3	+0.3	8.2	+0.3	2804	+327	2969	+220	17	0	13	+1
	2080	29.4		28.8		8.3		8.3		3042		2935		13		14	

The first is the CSIRO-Mk2 model from the Commonwealth Scientific and Industrial Research Organisation of Australia. The second is the HADCM3 model from the Hadley Centre for Climate Prediction and Research in the UK. It is uncertain how much GHG will be emitted in the future, yet the level of these emissions is likely to affect how trends in the climate develop. Taking into account some of the different emission scenarios is thus important in making reasonable predictions for Borneo's future climate. The IPCC has created different emission scenarios representing a range of development trajectories across the world. The following analyses use two of these scenarios that are commonly compared in climate modelling studies: a 'worst-case' scenario (entitled A2), and a 'best-case' scenario (B2). Under the socio-economic assumptions of 'worst-case', global development continues to be highly heterogeneous with the main driving forces being high levels of population growth, increased energy use, slow technological development and high land-use changes. The socio-economic future under 'best-case' represents a more sustainable globalized world, with slower population growth and land cover changes, and more diverse technological developments (Figure 6).

Temperature

The comparison of current and projected future climates reveals substantial climate change expected for Borneo this century (Figure 6). The clearest pattern is a projected rise in average annual temperature across the lowlands of Borneo. Island-wide mean temperature could rise by 2-3 °C above current levels by 2080. Some areas, particularly those in mountains, could see more extreme differences in temperature of around 5 °C (Figure 6). Coastal regions are predicted to show the most novel temperatures (Figure 7), which may be particularly evident in the island's urban areas. Borneo's major cities are predicted to warm by at least 2 °C, and temperatures of the administrative capitals of Banjarmasin (South Kalimantan), Kota Kinabalu (Sabah) and Palangkaraya (Central Kalimantan) are projected to rise by more than 10% of their current levels (Table 1). Projections are not just confined to average temperatures. Climate models also point to greater variability in future temperatures, for example as indicated by an increase in annual temperature range and seasonality over time (Figure 8). These changes are particularly pronounced in the HADCM3 climate model. Taking both these factors into account, the most affected region is predicted to be coastal East Kalimantan, from the Sangkulirang-Mangkalihat peninsula southwards to the cities of Samarinda and Balikpapan. This area currently experiences relatively stable temperatures throughout the year, but by the end of the century could face temperature fluctuations akin to the southeast and northwest parts of the island. Conversely, the lower catchment of the Kapuas river near to Pontianak in the west of Borneo, is expected to exhibit a lower range of temperatures over the year.

The projections of temperature are broadly similar across the majority of Borneo under the two emission scenarios (Figure 6). Outcomes are also broadly similar between the CSIRO and Hadley models, with the exception of the south-east of the island for which projections of temperature variation and seasonality increasingly diverge over time (Figure 6). For example, Hadley projections depict significantly greater temperature seasonality over Borneo than those from CSIRO models. This disparity might have implications for predicting the extent of orangutan habitat if distribution models are sensitive to these variables.

Rainfall

Over the course of this century Borneo is likely to experience elevated rainfall. However, the two climate models examined vary substantially in their projections. Under the CSIRO model projections, total annual rainfall over the Bornean lowlands could increase by up to 224-520 mm by 2050 and 513-754 mm by 2080, representing an increase of more than 15% of present levels in some parts of the island, including the large cities of Banjarmasin, Kota Kinabalu, Kuching, Palangkaraya and Samarinda (Figure 9; Table 1). However, the Hadley model predictions for rainfall are generally less severe, pointing to a marginal increase of, at most, 248-305 mm by the end of the century. Nevertheless, these figures mask subtle differences in rainfall patterns across the island, most notable of which is that some regions, such as the southwest, may experience little change, or potentially even see reduced rainfall in the future at certain times of the year (Figure 9). This pattern in the southwest in fact continues the decline in precipitation that has already occurred between 1850 and 1980 in this region, as shown in Figure 5. The projected variation in annual rainfall is mirrored by considerable projected changes to seasonality. Currently the greatest rainfall seasonality is experienced along the coastlines of western Sarawak, northern Sabah and South Kalimantan, as well as the headwaters of the Kapuas River in West Kalimantan. The future projections indicate the northernmost regions could experience less seasonality. Highly novel patterns of total rainfall are predicted in the Usun Apau mountain range straddling the border of Sarawak and North Kalimantan (Figure 10). This area also sees the greatest difference in climate projections under the two emission storylines – predictions of rainfall during the driest month vary widely between the worst-case and best-case scenarios in this region, but elsewhere on Borneo are very similar and appear largely consistent over time.

COMPARISON OF POTENTIAL CLIMATES FOR BORNEO

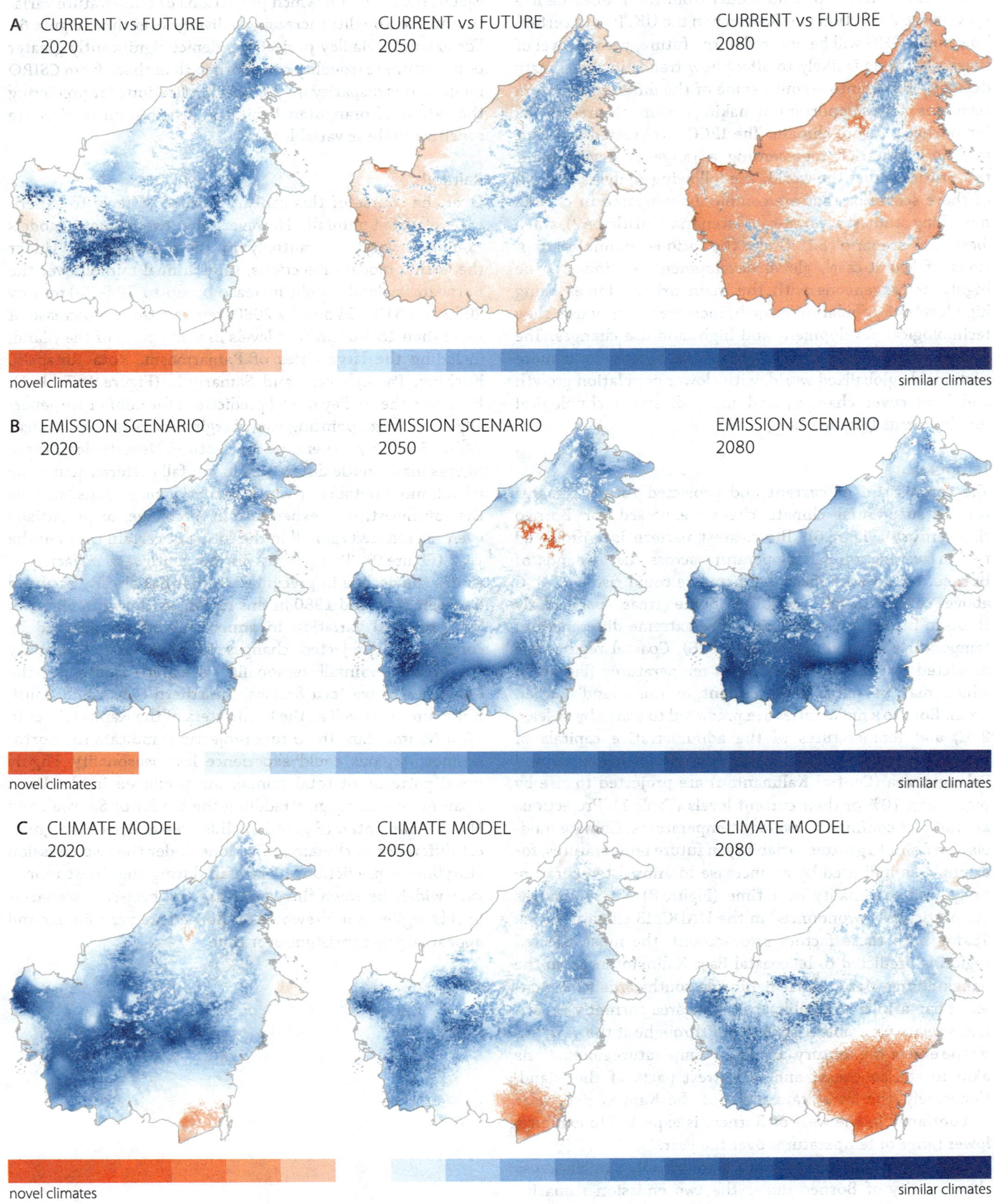

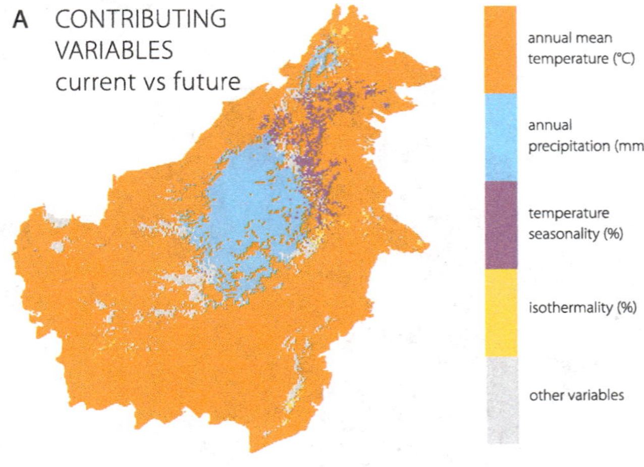

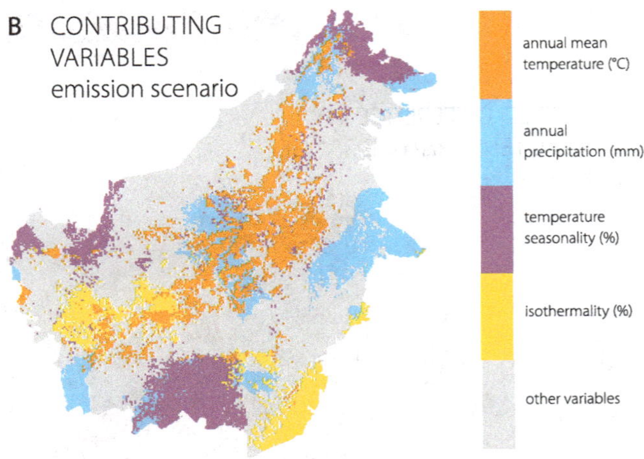

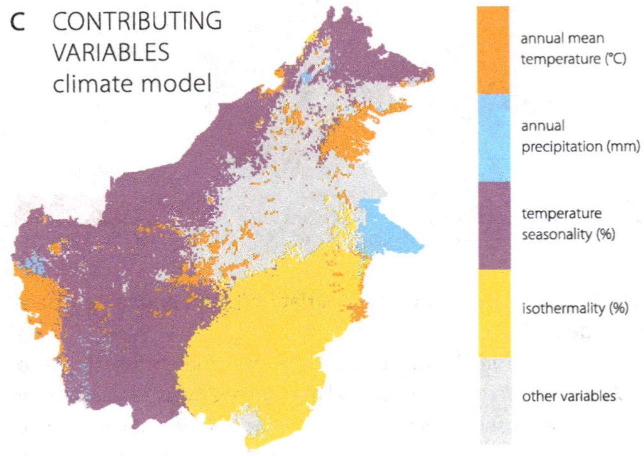

Figure 6 **Comparison of potential climates for Borneo, using a subset of eight bioclimatic variables and multivariate environmental similarity surfaces (MESS).**

Areas of the island with distinct, or novel, climates in each time period are indicated by negative values (red), while areas with similar climates are indicated by positive values (blue). Pairwise comparisons are made between current and future climates (upper panel) using the CSIRO worst-case scenario, as well as emission scenario (middle panel for CSIRO best-case versus worst-case scenario) and climate projection model (lower panel for CSIRO versus Hadley under a worst-case emission scenario). Accompanying each MESS is a map of contributing variables (otherwise known as 'limiting factors', right pane), which shows the variable that most influences model prediction at any given point. Cross referencing areas of novel climates with corresponding areas on these maps allows us to identify the most important variable(s) contributing to climate novelty. For clarity only the top four contributing variables are shown (grey areas indicate any of the four remaining variables).

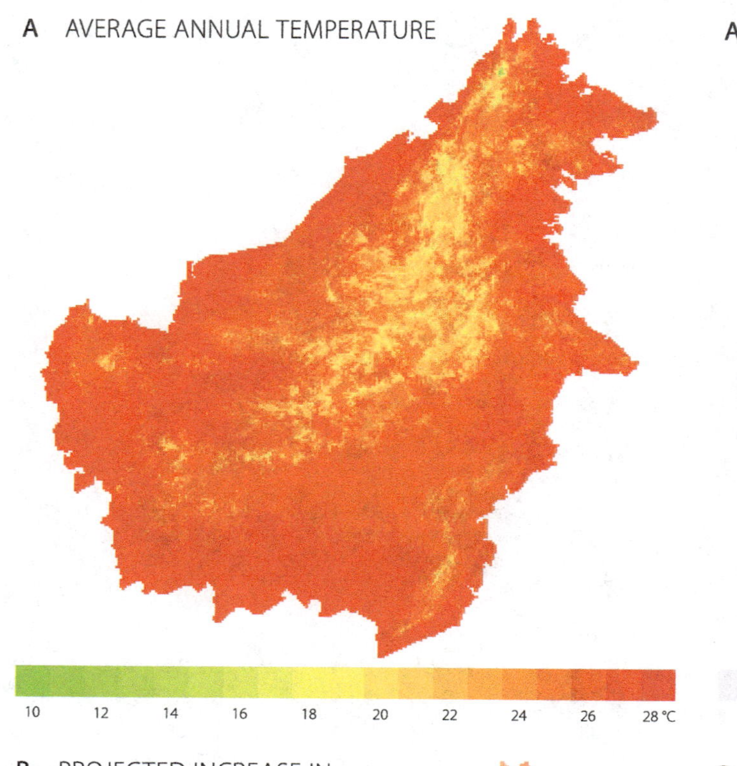

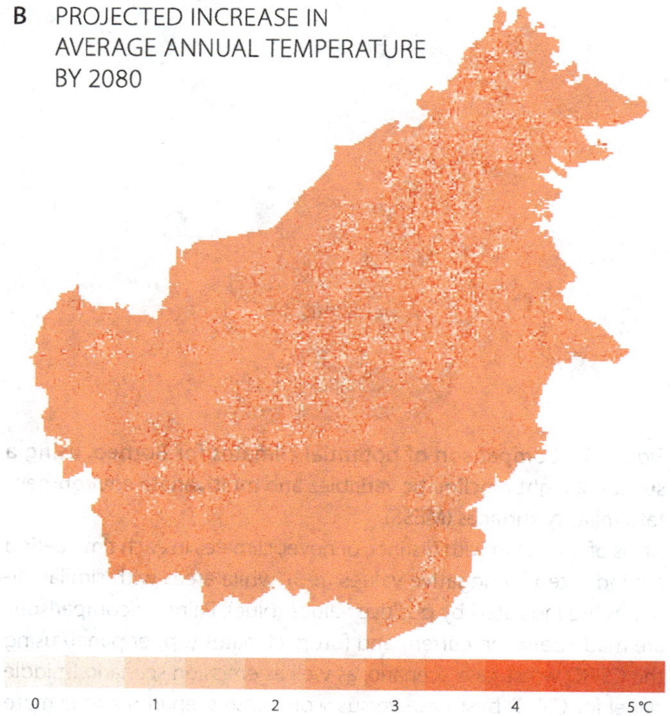

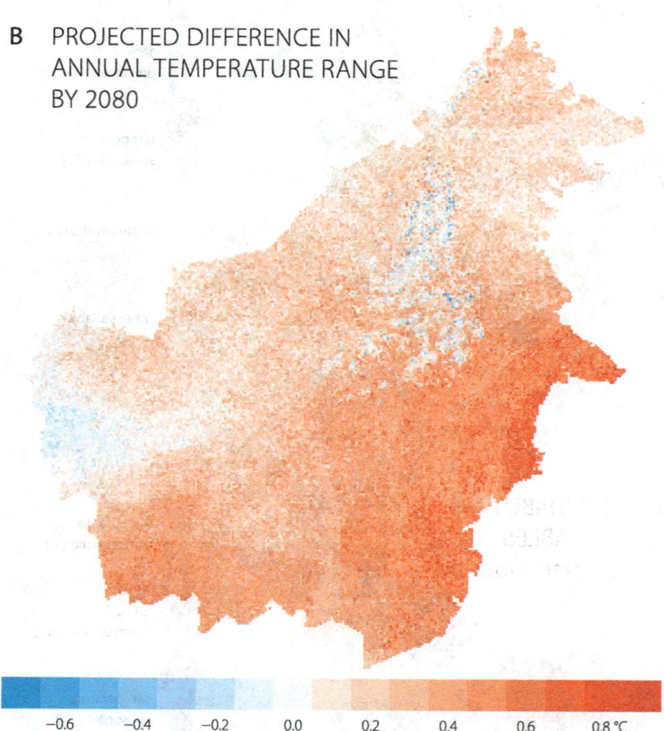

Figure 7   Current and projected variation in projected annual mean temperature for Borneo.
Average annual temperature under current climate conditions is greatest in the lowlands (A), which are projected to experience at least 2 °C higher temperatures by 2080 according to the CSIRO climate model, worst-case emission scenario (B).

Figure 8   Current and projected variation in the annual temperature range for Borneo.
Annual temperature fluctuations are lowest in the north and east of the island, and greatest in South Kalimantan and west Sarawak under current climate conditions (A). According to CSIRO worst-case climate projections, the east of Borneo could see more stable temperature fluctuations by 2080, while the Pontianak and Kubu Raya districts of West Kalimantan could face greater temperature seasonality (B).

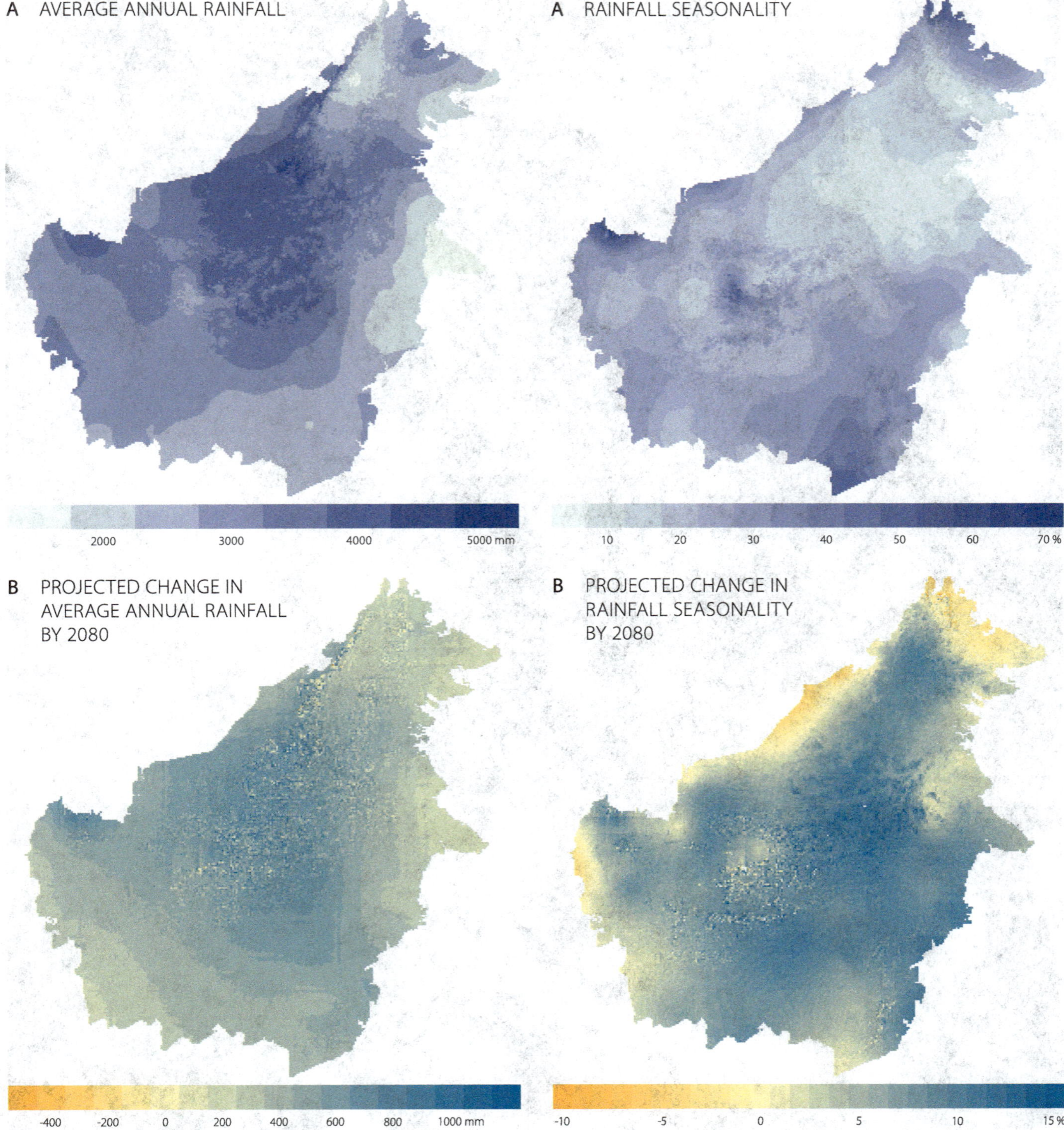

Figure 9 **Current and projected variation in annual precipitation for Borneo.**
The areas of lowest annual rainfall (<2 500 mm per year) are currently in the east, south and northeast, and the highest rainfall (>3 200 mm per year) is experienced in western Sarawak and the central mountains (A). These latter areas could face greater rainfall by 2080 according to CSIRO 'worst-case' climate projection (B).

Figure 10 **Current and projected seasonality in rainfall for Borneo.**
Northern Sabah, western Sarawak and South Kalimantan currently experience the greatest seasonality in rainfall (A). According to CSIRO worst-case climate projections reduced seasonality is to be expected in Brunei, northern Sabah and coastal West Kalimantan by 2080, while interior areas may experience increased seasonality (B).

Unflanged male, Tuanan, Central Kalimantan.

Photo Perry van Duijnhoven

Mother with infant, Tuanan, Central Kalimantan. **Photo** Perry van Duijnhoven

Loading Illegaly cut logs on the back of a truck, Danau Sentarum, West Kalimantan.

Photo Perry van Duijnhoven

LAND-COVER AND CLIMATE CHANGE IMPACTS ON THE BORNEAN ORANGUTAN

Tropical species and their association with climate

Climate change and habitat alteration are two of the greatest anthropogenic threats faced by terrestrial biodiversity, and so represent major challenges and priorities in conservation (BROOK et al. 2008). These two threats acting together may prove particularly difficult for the survival of many species, including the orangutan. Across the world the ranges of numerous species are shifting to higher latitudes or elevations in association with climate warming, and local extinctions continue to be documented for those species that have been unable to do so (PARMESAN 2006). Recent estimates indicate that, on average, every ten years range-shifting species are moving northwards or southwards by around 17 km, and upslope by 11 m (CHEN et al. 2011). As a result of this shift, climate change is playing an increasingly large role in the survival of species. Its impact in combination with land-cover change also remains important to determine. Land cover changes continue to cause population declines and extinctions as suitable habitat is lost, fragmented and degraded (IUCN 2013). It is then reasonable to imagine that losses and alteration of important habitats will have a stronger influence on species before the effects of climate change are felt – especially for many tropical species that inhabit regions experiencing high rates of deforestation such as Southeast Asia (JETZ et al. 2007; SALA et al. 2000). There is now a growing realization amongst conservation scientists that both of these environmental change processes may act together, potentially creating a double jeopardy for many species. Identifying which species may be affected in this way is clearly important for conservation planning. Although climate changes might be less severe in the tropics compared to vast temperate regions such as Europe and North America, it is limitations to distributional ranges and long-distance movements that could enhance vulnerability of tropical wildlife (TEWKSBURY et al. 2008). The tropics typically have narrow latitudinal temperature gradients compared to temperate areas, and so climate-induced movements of species to higher elevations are considered more likely than latitudinal range shifts (COLWELL et al. 2008). In fact, such range shifts have already been documented on Borneo: on Mount Kinabalu in Sabah scientists reported upslope range movements of moths over a 42 year period (CHEN et al. 2009), and many species exhibited narrowing altitudinal ranges over time (MEIJAARD & SHEIL 2008).

The potential responses of tropical species to climate and habitat changes can be summarized by three general predictions (COLWELL et al. 2008; SVENNING & CONDIT 2008). First, tropical lowlands (typically land below 500 m in elevation) may experience a process termed 'biotic attrition': warming drives species out of the lowlands, but no source of species adapted to higher temperatures is available to compensate for the losses. The other two predicted risks for tropical mountain species are range-shift gaps (where species' current altitudinal ranges do not overlap climatically with suitable ranges of the future) and mountain-top extinction (where warming pushes climatically suitable conditions off mountain peaks). Each result is made more complicated when considering whether sufficient habitat, for example tropical rainforest, will be available for wildlife to move through and into. Given potential interactions between temperature shifts and habitat suitability, it is clearly important that assessments of climate change impacts on threatened species should also consider the effects of land cover changes.

The Bornean orangutan distribution

As a tropical species, the Bornean orangutan may be particularly sensitive to the combination of these multiple threats. Estimating the response of orangutan distribution to future climate and land-cover change is not only important for the conservation of this iconic species, but it may also indicate how the broader biodiversity patterns in the region will be affected since orangutans are a useful 'indicator' species. Orangutans are among the few Bornean species for which detailed distribution maps are available based on extensive field surveys. Orangutans are also ecologically of medium sensitivity, generally dependent on forests, but not restricted to very narrow ecological niches within those forests. Orangutans therefore represent a relatively large group of forest species that inhabit Borneo, including other iconic mammals, which raises the following questions: 1.) How the climate change projections are expected to affect the habitat suitability for orangutans; 2) How predicted land-cover change alone affects the distribution of the Bornean orangutan; and 3) How these projections combine.

Until relatively recently, assessments of orangutan distribution in the wild were constrained by the availability of geographically referenced field surveys. As such they were based on broad summaries of the areas in which these animals could be found. The latest update for the Bornean orangutan (WICH et al. 2012), however, has capitalized on new field surveys and several developments in species distribution modeling in order to project the species range and identify suitable habitats. The current core distribution of orangutans on Borneo was predicted to be around 155 000 km² in forested

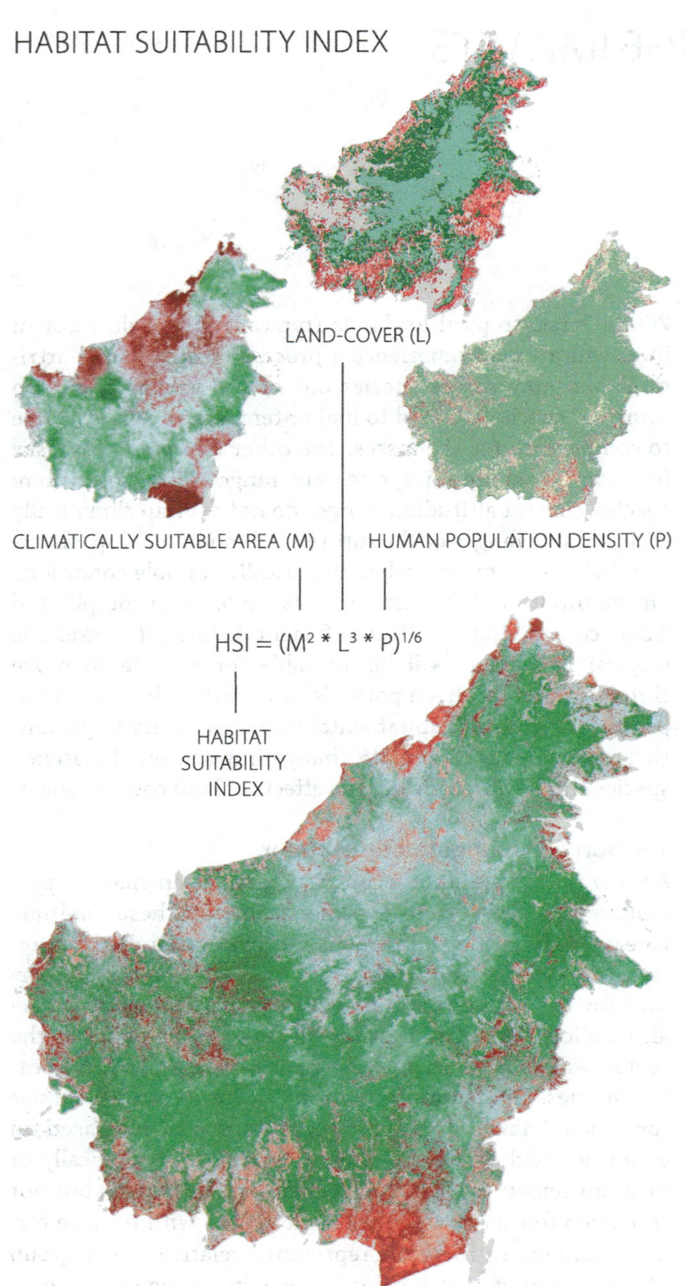

$$HSI = (M^2 * L^3 * P)^{1/6}$$

areas, and was determined by mapping all known localities where the species has been found present (1990-2011), and associating these points with bioclimatic and land cover variables (STRUEBIG *et al.* 2015).

The effects of climate change on the future orangutan distribution

By using bioclimatic modeling to map present-day 'climatically suitable areas' for the orangutan, it is possible to project their potential distribution into the future. One can deduce the general climates that might be suitable for a species, such as orangutans, because climate conditions like temperature and rainfall patterns often define the places that those species can occur. The eventual suitability of an area for a species is also defined by land cover (*e.g.* lowland forest, mangrove, old plantations) and potential threats from humans. These model outputs are most informative when raw values are presented, ranging from low to high suitability. To derive area estimates of suitable habitat, however, a conservative threshold that allows for up to 10 percent error in model predictions is required below which land is deemed unsuitable. Using a threshold of this type could lead to overprediction, rather than underprediction, of suitable areas for the orangutan. Under current conditions 260 000 km² of habitat within the known core orangutan range is deemed potentially suitable. Based on land cover maps estimated for the 1950s, 316 000 km² would have been suitable under the same climate conditions.

The model predicting current habitat suitability for orangutans can then be used to predict the future patterns under the differing climate GCMs for 2020, 2050 and 2080 that were presented in the previous chapter. In the analyses the environmental parameters that have the greatest predictive power for orangutan presence are: 1) annual temperature range, 2) precipitation in the driest month and quarter, and 3) mean diurnal temperature range. Notably, not all the areas that are considered suitable for orangutans currently host populations of the species (Figure 12), but they might provide viable refuge in the future.

Figure 11 **Schematic diagram illustrating the calculation of a habitat suitability (HSI) index for orangutan.**
The potential species distribution under each environmental change scenario is reflected by areas of suitable habitat. This is calculated by constraining the climate-driven probability of occurrence (M) by information on habitat affinity (L) and sensitivity to human population pressure (P), scores which were provided by four Bornean orangutan experts. Dark green areas are highly suitable; dark red areas are highly unsuitable.

Figure 13 (page 34-35) **Forecasts of habitat suitability for Bornean orangutan when accounting for the effects of climate change only.** Each map summarizes habitat suitability based on climate and land cover data. Climate data are derived from one of two climate models (CSIRO or Hadley) projected under one of two emission scenarios (best and worst scenarios). Land-cover data are fixed to 2010 levels. See Figure 11 for an explanation of how maps habitat suitability was calculated. Color gradient adjusted to visually improve contrast between high and low suitability areas in future time periods.

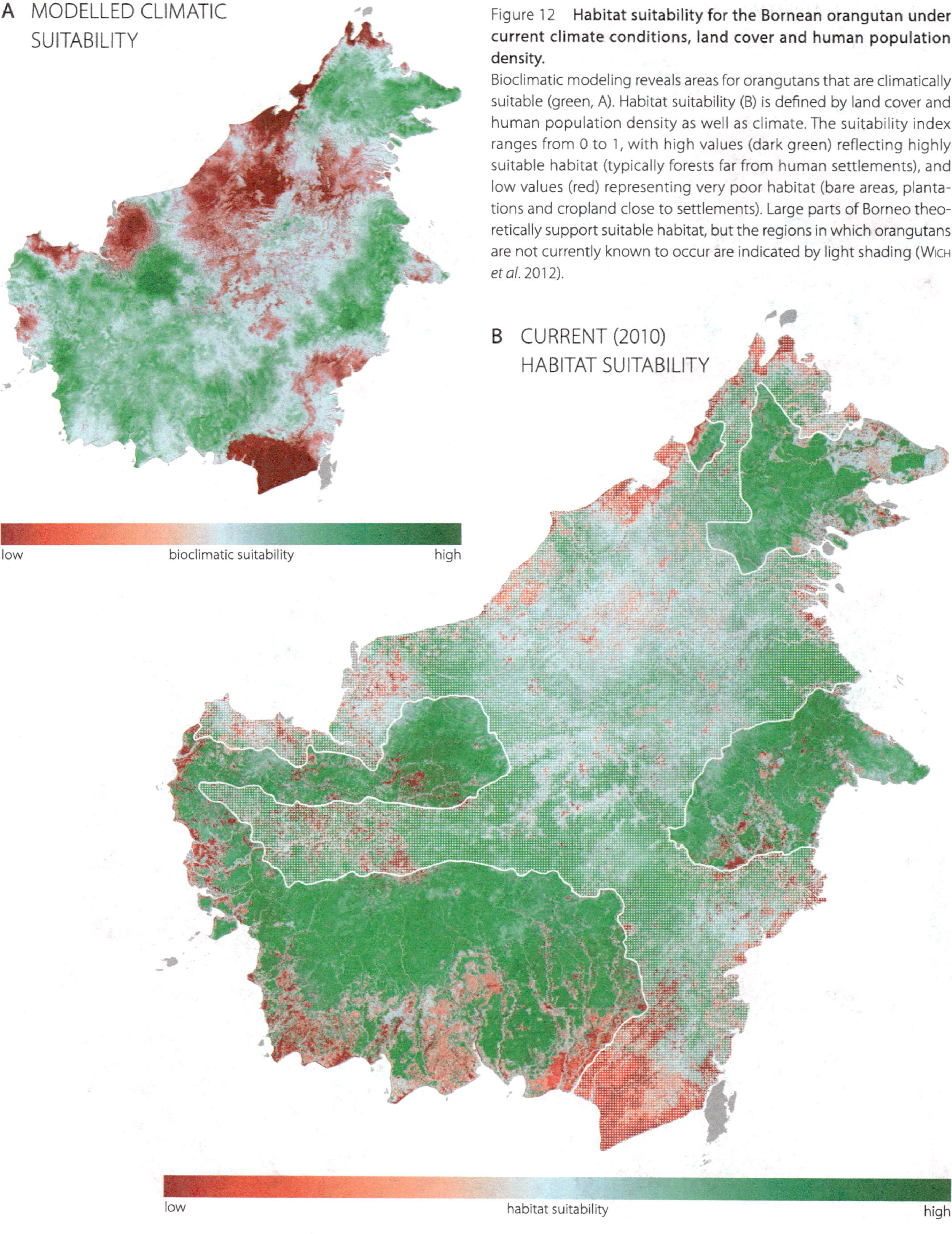

Figure 12 **Habitat suitability for the Bornean orangutan under current climate conditions, land cover and human population density.**
Bioclimatic modeling reveals areas for orangutans that are climatically suitable (green, A). Habitat suitability (B) is defined by land cover and human population density as well as climate. The suitability index ranges from 0 to 1, with high values (dark green) reflecting highly suitable habitat (typically forests far from human settlements), and low values (red) representing very poor habitat (bare areas, plantations and cropland close to settlements). Large parts of Borneo theoretically support suitable habitat, but the regions in which orangutans are not currently known to occur are indicated by light shading (WICH et al. 2012).

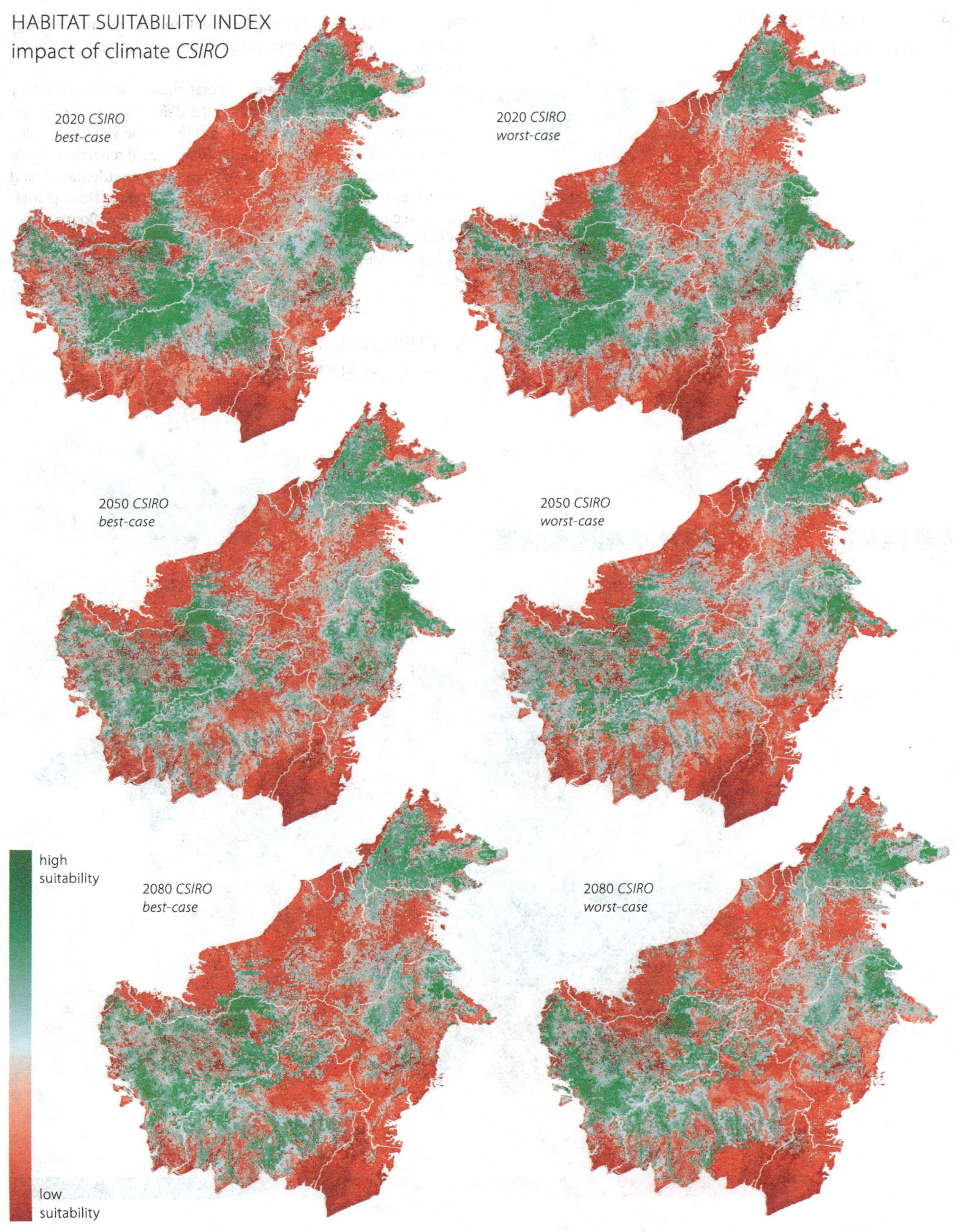

HABITAT SUITABILITY INDEX
impact of climate CSIRO
2020 CSIRO best-case
2020 CSIRO worst-case
2050 CSIRO best-case
2050 CSIRO worst-case
2080 CSIRO best-case
2080 CSIRO worst-case
high suitability
low suitability

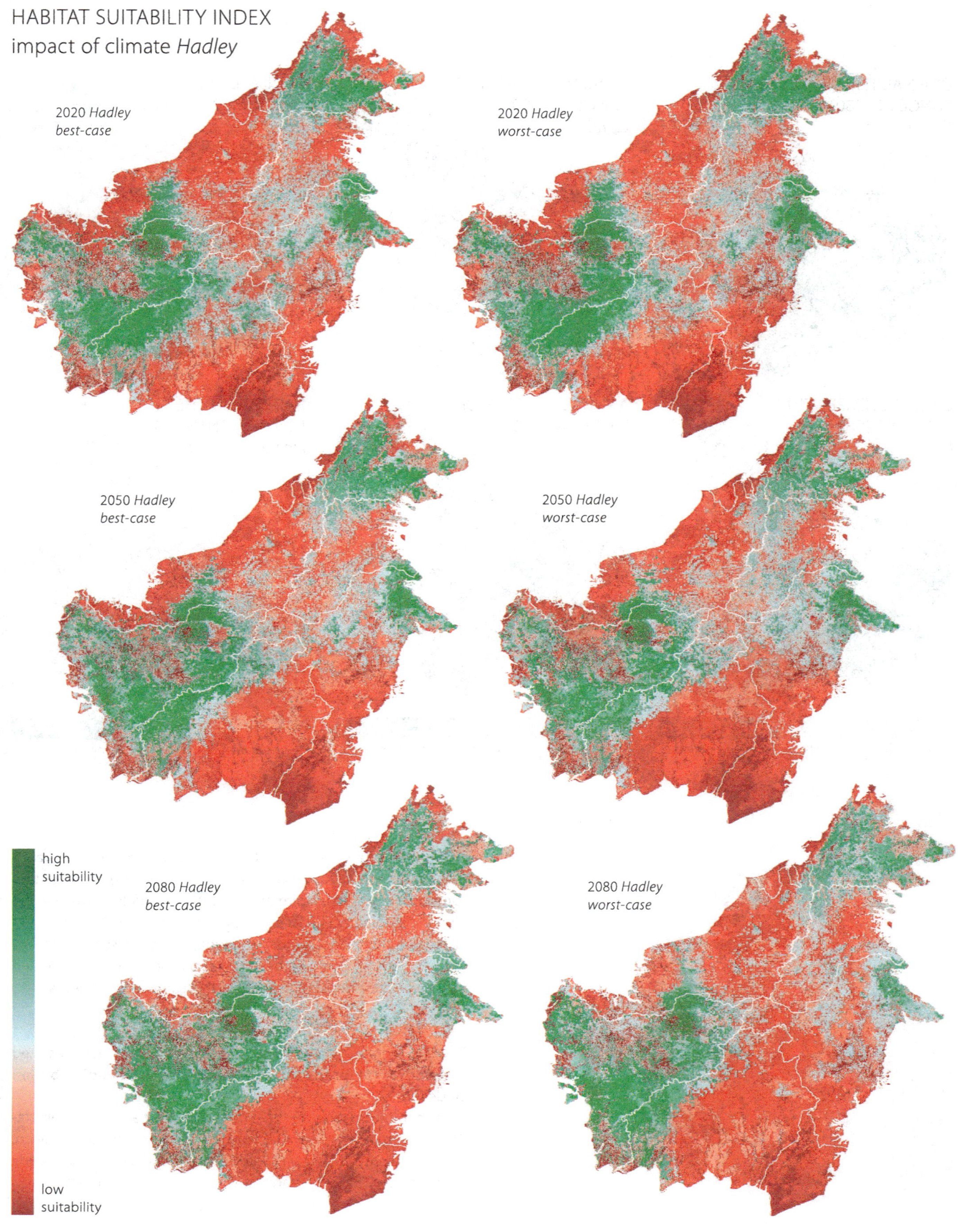

HABITAT SUITABILITY INDEX
impact of climate Hadley
2020 Hadley best-case
2020 Hadley worst-case
2050 Hadley best-case
2050 Hadley worst-case
2080 Hadley best-case
2080 Hadley worst-case
high suitability
low suitability

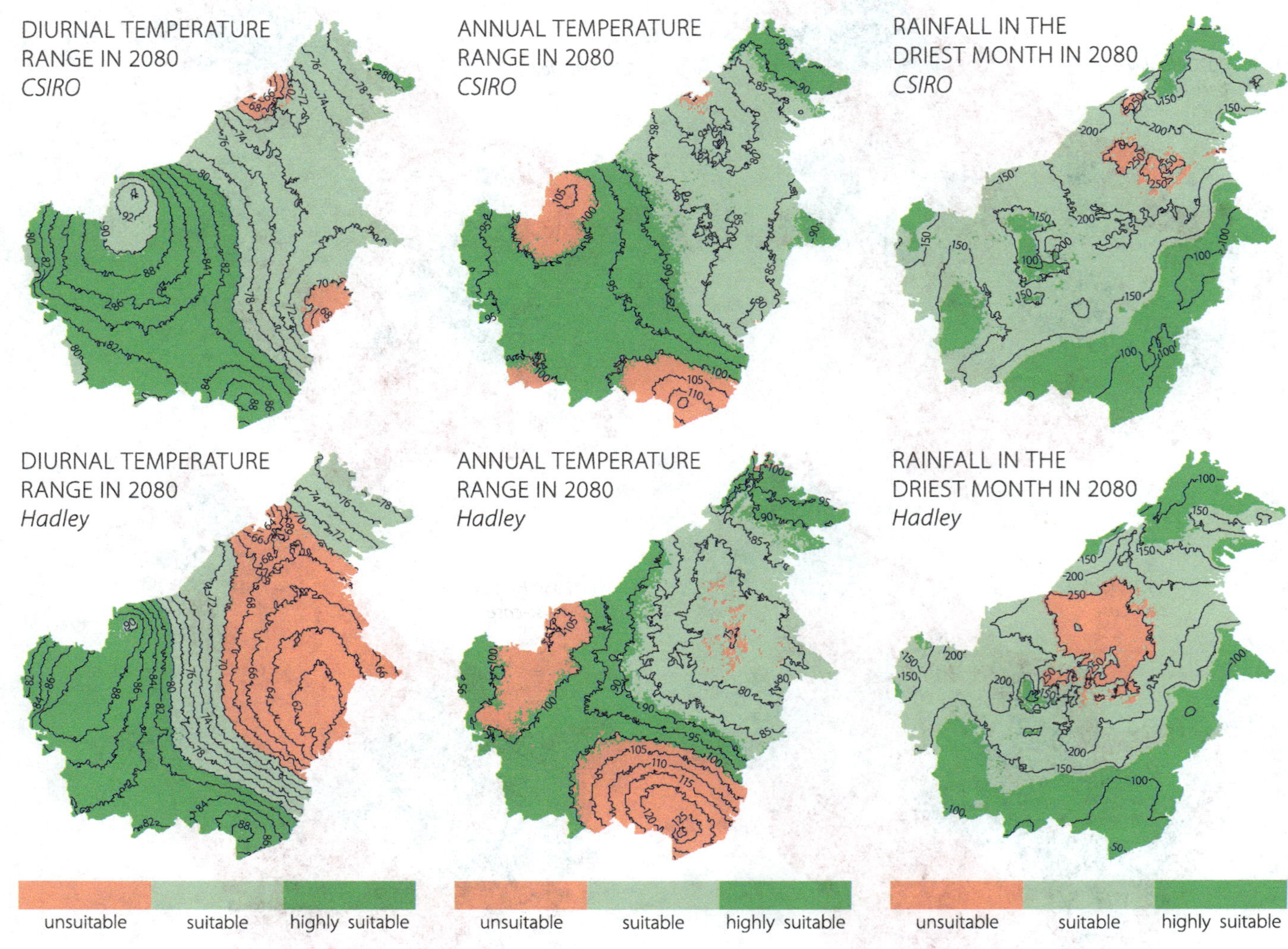

Figure 14   **Differences in bioclimatically suitable areas for orang-utan on Borneo in 2080 as defined by three climate variables that contribute highly to bioclimatic distribution models for the species.** Climatic suitability is based on thresholds from response curves of bioclimatic models under the worst-case emission scenario. A larger extent of Borneo is deemed 'unsuitable' by models based on Hadley climate forecasts compared to those from CSIRO forecasts.

Figure 15 (right page)   **Land-use allocation 2010 (A) and predicted forest cover (B) for the future time periods used in distribution modeling.** Note that information on the extent of industrial oil-palm plantations in Sabah was unavailable when models were undertaken, but the majority of land presented here for 'conversion' has already been planted. This discrepancy does not affect the outcome of this appraisal.

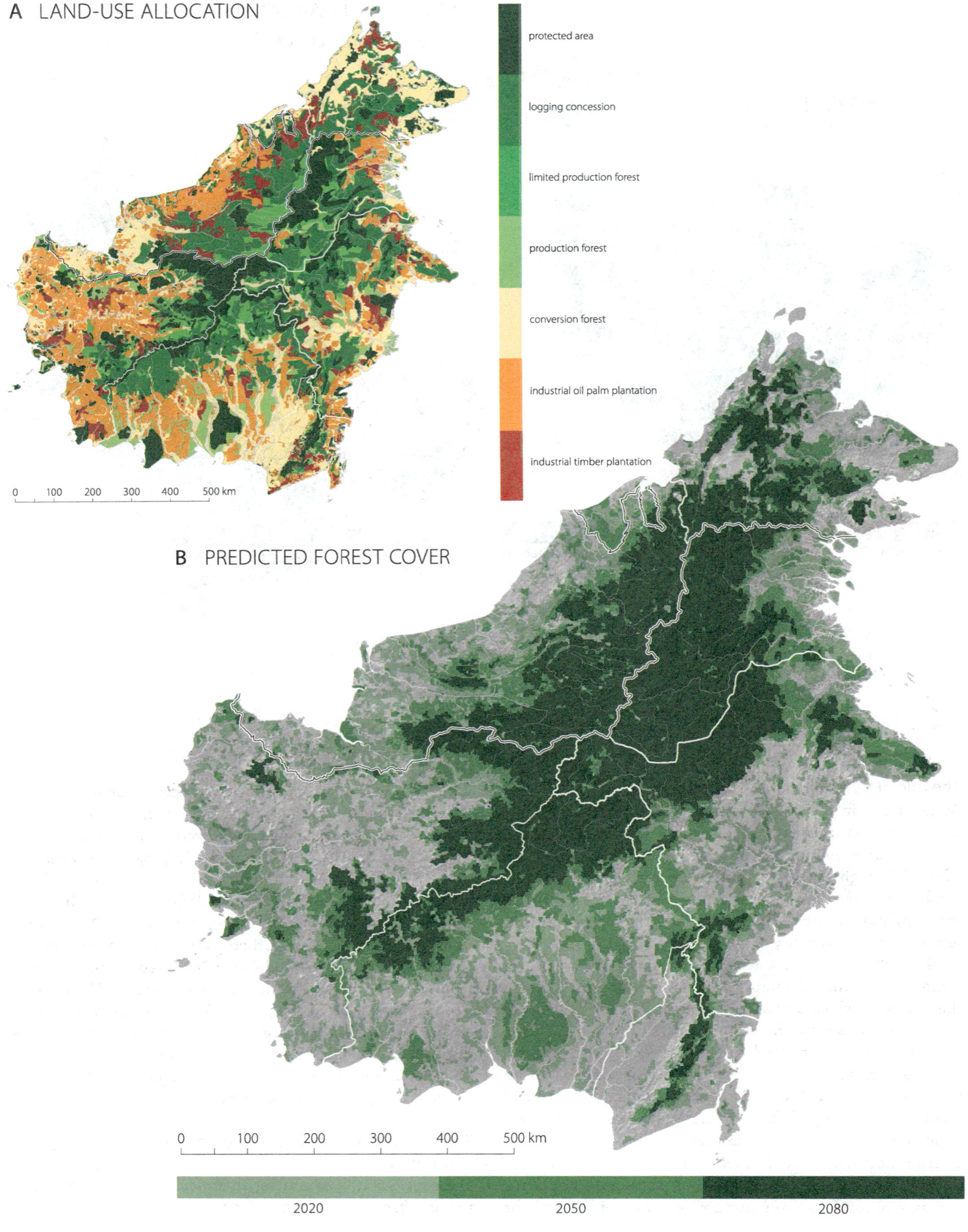
A LAND-USE ALLOCATION
protected area
logging concession
limited production forest
production forest
conversion forest
industrial oil palm plantation
industrial timber plantation
0 100 200 300 400 500 km
B PREDICTED FOREST COVER
0 100 200 300 400 500 km
2020
2050
2080

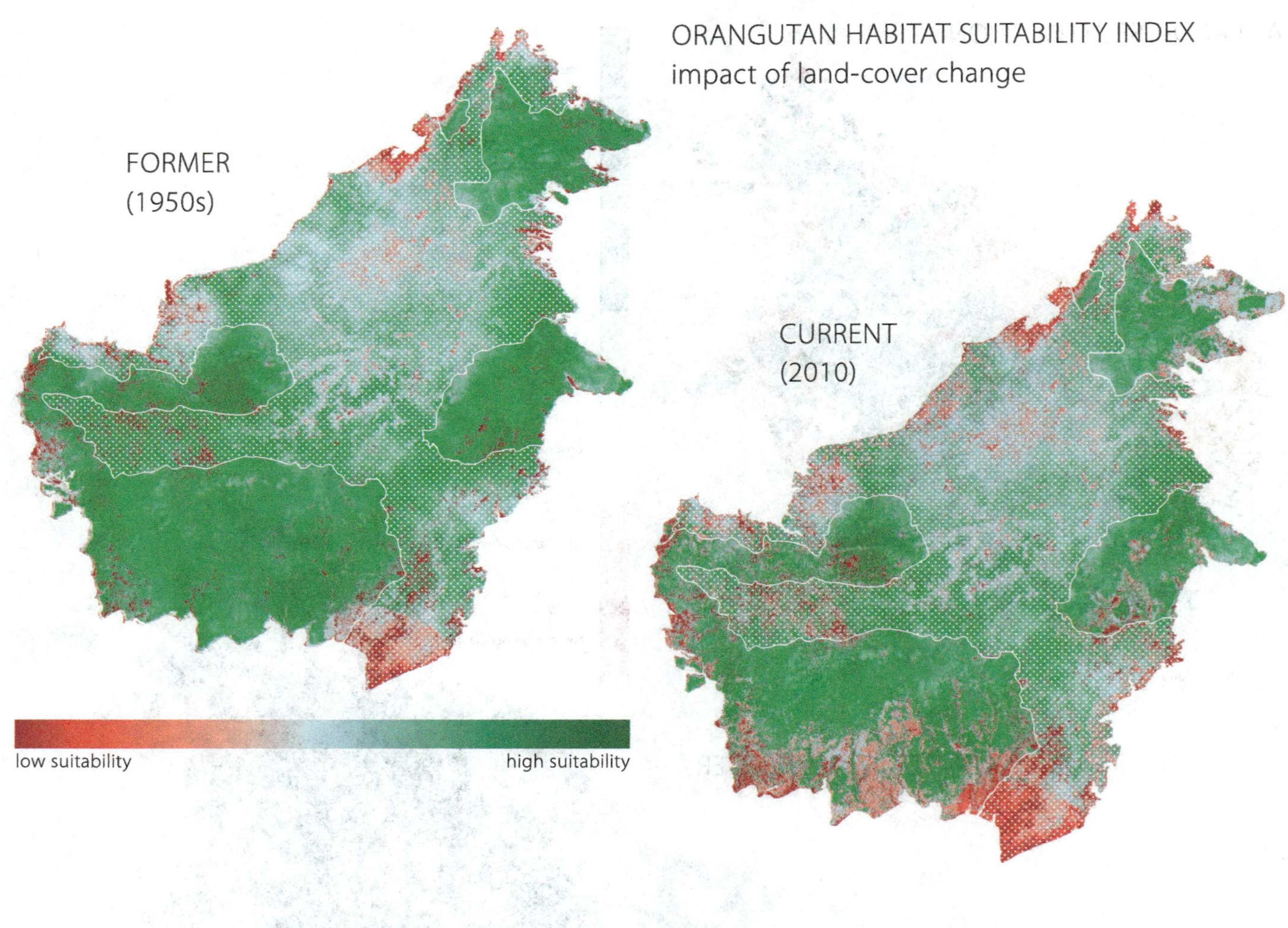

The role of different climate projections

As described in the section on climate change, the climate predictions were made via two different GCMs and two different emission scenarios for the time periods 2020, 2050 and 2080. Although the projections from each model-scenario combination result in similar patterns of habitat suitability for orangutans which declines over time, there are subtle differences in the trajectory between models (Figure 13).

These differences as a result of a small increase in, for example, the annual temperature range, may just exceed the limits by which suitability for that measure is defined. The affected areas are therefore predicted to be 'unsuitable' for orangutans as opposed to suitable. For the most part, there is a broad agreement between the models based on the two GCMs, but the Hadley model predictions result in models with lower probability of orangutan presence in some regions, in particular in the southeast of Borneo and in the interior northwards to Brunei. These areas coincide with marked differences between the GCM projections in diurnal and annual temperature range between Brunei bay and the Mahakam river delta, and also with precipitation in the driest month across the south of the island (Figure 14). Within the known core range of orangutans a gradual decline in suitable habitat over time is consistent under all projections assessed, with 130 000-205 000 km² predicted to remain by 2050. By 2080 79 000-130 000 km² of suitable habitat is predicted to remain under CSIRO projections, but 82 000-111 000 km² under Hadley projections. There is little difference between the habitat suitability model outputs based on climate data from the two emission scenarios, 'worst-case' and 'best-case'. The similar output in the two models indicates that climate change is expected to have a marked impact, regardless of the development trends of regions or nations. Concerted efforts between nations to cut emissions at the global level may still improve this pessimistic outlook.

The effects of land-cover change on the future orangutan distribution

The conversion of rainforest to agricultural plantations has major effects on tropical biodiversity and on orangutans

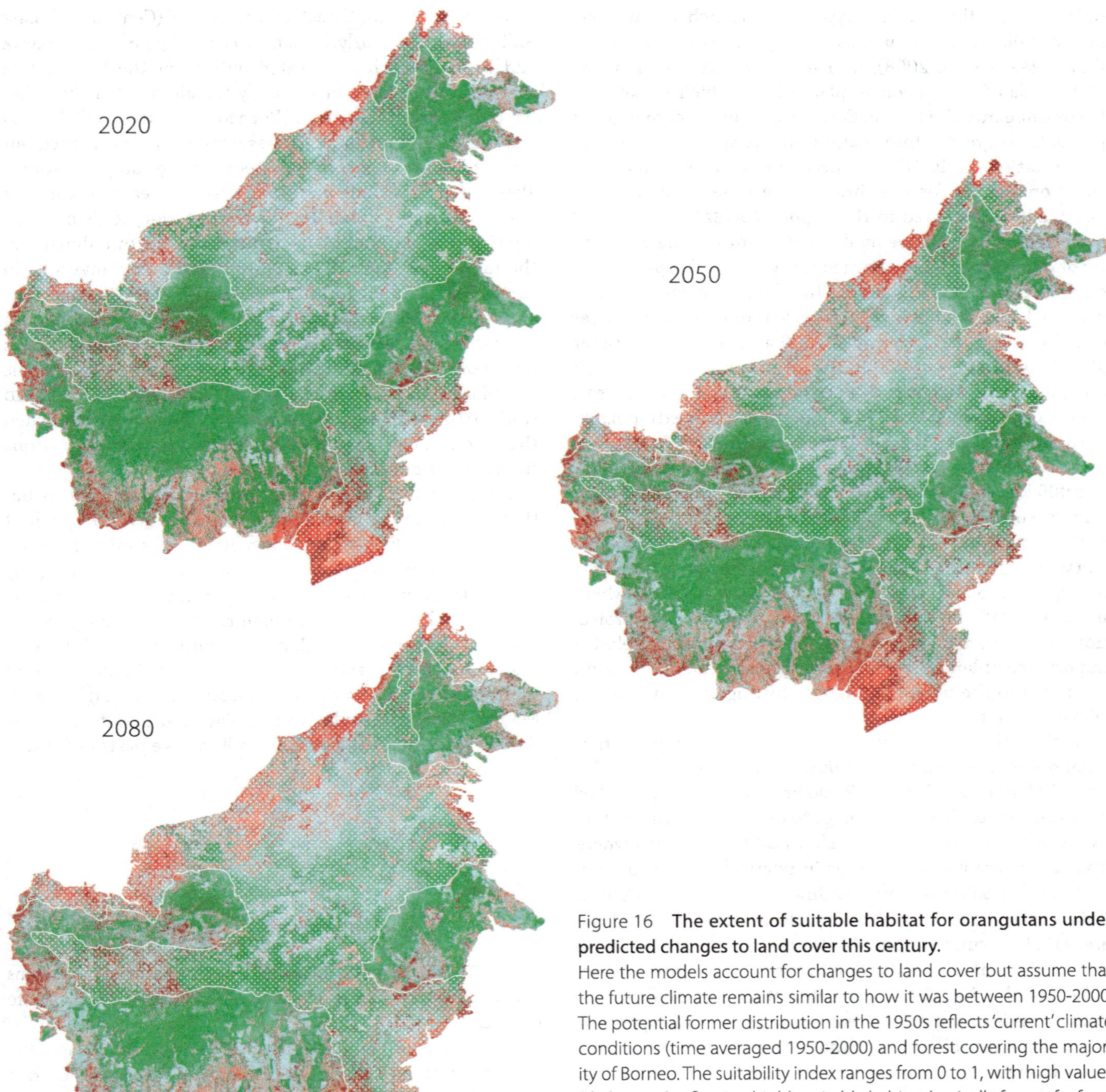

Figure 16 **The extent of suitable habitat for orangutans under predicted changes to land cover this century.**

Here the models account for changes to land cover but assume that the future climate remains similar to how it was between 1950-2000. The potential former distribution in the 1950s reflects 'current' climate conditions (time averaged 1950-2000) and forest covering the majority of Borneo. The suitability index ranges from 0 to 1, with high values (dark green) reflecting highly suitable habitat (typically forests far from human settlements), and low values (red) representing very poor habitat (bare areas, plantations and cropland close to settlements). Most reduction in habitat suitability is predicted to occur in the lowlands of West, Central and East Kalimantan, reflecting high projected deforestation in these provinces. Nevertheless, the majority of habitat that is currently suitable would be predicted to remain suitable in the future if climate change was not to occur. Large parts of Borneo theoretically support suitable habitat, but the regions in which orangutans are not currently known to occur are indicated by light shading (Wich et al. 2012). See Figure 11 for an explanation of how habitat suitability was calculated.

in particular. These effects typically outweigh the impacts of other disturbances such as logging (GIBSON *et al.* 2011; FITZHERBERT *et al.* 2008). Borneo is currently experiencing high levels of conversion to plantations, which is expected to continue into the future. Conversion will affect orangutan populations greatly. Information on the spatial patterns of deforestation on the island is needed to reliably predict land cover on Borneo for the three future time periods (2020, 2050, 2080) presented in this report. GAVEAU *et al.* (2014) have created a predictive model that quantifies the probability of deforestation over Borneo using fine-scale spatial data on forest loss in Kalimantan. The deforestation base-map used to develop the model portrays loss of natural tree cover over a ten year period between 2000 and 2010 (CARLSON *et al.* 2013; GAVEAU *et al.* 2013). The mean annual deforestation rate for Kalimantan over this period was 2 341 km² per year, which corresponded to 3 234 km² per year when extrapolated to the entire area of Borneo. Assuming a similar deforestation rate in the future, 32 000 km² of forest could be lost by 2020, 129 000 km² by 2050, and 226 000 km² by 2080 (Figure 15) – representing approximately 7%, 28% and 49% of the current forest cover in 2010. These results could be considered a worst-case scenario projection, as deforestation rates might be expected to slow over time as they have done elsewhere in Asia (KOH 2007). However, such a pessimistic projection, however, allows for a better evaluation of the relative importance of land cover on limiting orangutan distribution compared to the potentially range-shifting consequences of climate change.

Within the known current core range this projection amounts to a reduction of habitat to 251 000 km² by 2020, 237 000 km² by 2050 and 219 000 km² by 2080. This level of deforestation could be a challenge to orangutan conservation as much of this area corresponds to parts of Borneo where orangutans are known to occur in relatively high densities, such as the Sabangau and Katingan peatlands in Central Kalimantan (HUSSON *et al.* 2009). Nevertheless, most of the areas that are currently suitable for orangutans are predicted to remain suitable under land-cover change projections, reflecting the fact that most deforestation is predicted to occur outside of protected areas in degraded landscapes that are easily accessible – places that currently do not support many orangutans.

A precarious combination

Combining the projections for land cover change with those for climate change arguably provides the most realistic image of the potential future orangutan distribution. Worryingly, this combination reveals a much lower extent of highly suitable habitat than models incorporating land cover alone (Figure 17). Considering both a changing climate and land cover the area of suitable habitat within the orangutan core range is also expected to decline over time. Model estimates indicate that 110 000-165 000 km² of land would be suitable by 2050, and 49 000-83 000 km² by 2080. Most of the projected habitat loss in the core orangutan range is projected

to occur in the southeast of the island (Central and East Kalimantan). Similarly, lowland areas in the north of Sarawak and Brunei, which are located outside of the known core range, could become increasingly unsuitable. On the other hand there are other parts of Borneo which are highlighted as suitable habitat under all the assessments in this report, but are not known to currently support orangutan populations. These regions include western Sabah and eastern Sarawak (Lawas division) extending to the interior of Temburong district in Brunei and Malinau district in North Kalimantan; the interior of Kapit, Sibu, Bintulu and Sarike divisions in Sarawak, around the forested headwaters of the Rejang river; and the foothills of the Madi plateau in Sintang and Kapuas Hulu districts. The reason for the absence of orangutans is not known, but could be due to past hunting or geographic boundaries such as rivers limiting movements (MEIJAARD *et al.* 2010). If threats to orangutans can be controlled then these areas might provide suitable sites for reintroductions from captive populations elsewhere.

An estimate of the potential future orangutan distribution would use the current extent and assess how much of this range is likely to remain suitable. Notably, there are some important regions where orangutans are currently known to occur which are not expected to be negatively affected by land cover or climate changes according to this analysis. These regions include the foothills of the Schwaner mountains in West and Central Kalimantan, Kapuas lakes of West Kalimantan near the Sarawak border, the Sangkulirang-Mangkalihat peninsula and Kelai catchment of East Kalimantan (north of Kutai), as well as large parts of Sabah.

Figure 17 **Forecasts of habitat suitability for Bornean orangutans when accounting for the effects of both land cover and climate change.**

Each map summarizes habitat suitability based on climate and land cover data that change over time. Climate data are derived from one of two climate models (CSIRO or Hadley) projected under a best-case emission scenario, and land cover data are informed by a predictive deforestation model developed for the island. See Figure 11 for an explanation of how habitat suitability was calculated. Major changes to the extent of suitable habitat over time are predicted under all changing climate projections. The most affected areas are the northern and southeastern part of the island. Strongholds of suitable habitat are predicted to remain for substantial parts of the current orangutan range: namely forests in the interior of West Kalimantan, East Kalimantan and Sabah. Color gradient adjusted to visually improve contrast between high and low suitability areas in future time periods. The regions in which orangutans are not currently known to occur are indicated by light shading (WICH *et al.* 2012).

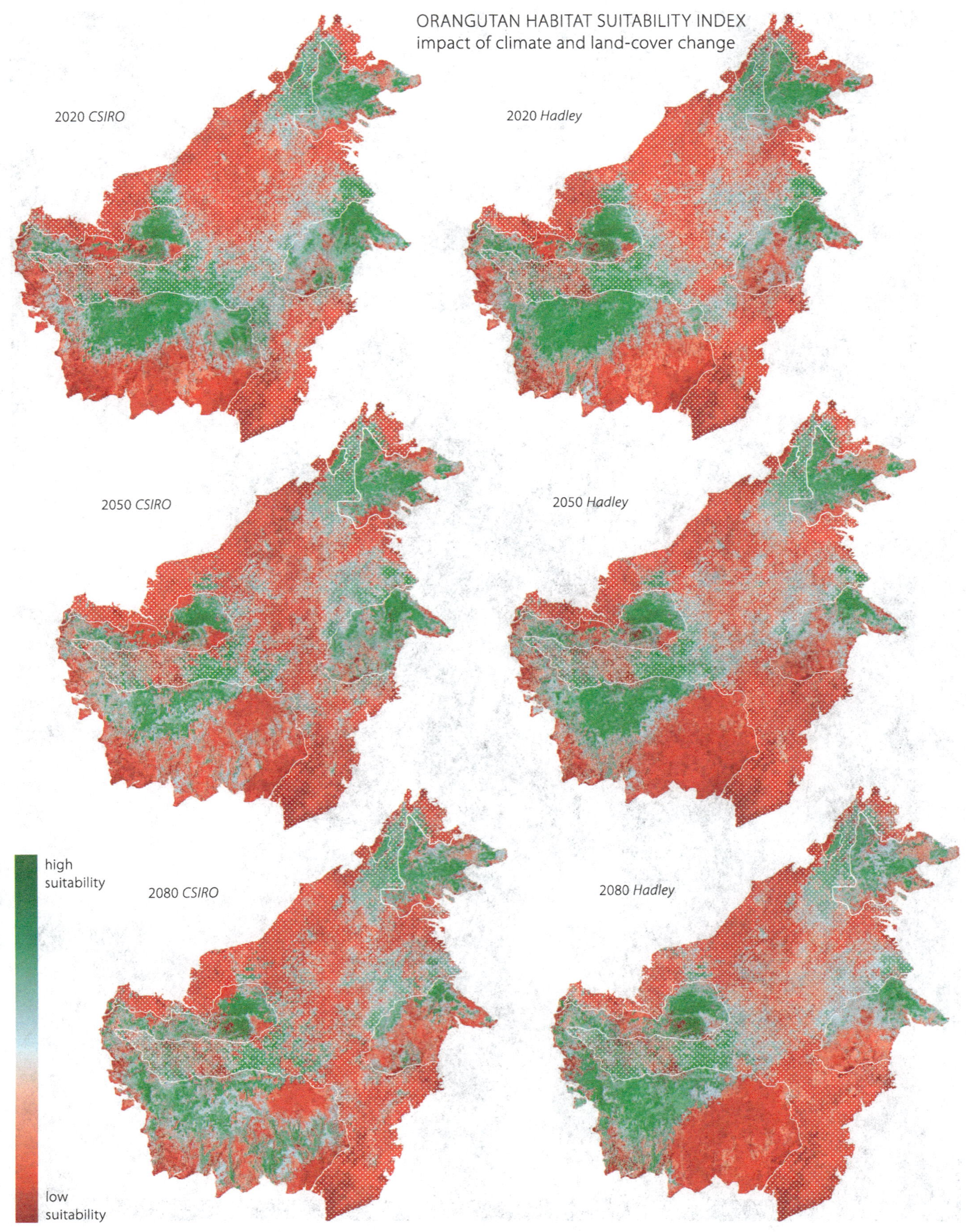

ORANGUTAN HABITAT SUITABILITY INDEX
impact of climate and land-cover change
2020 CSIRO
2020 Hadley
2050 CSIRO
2050 Hadley
2080 CSIRO
2080 Hadley
high suitability
low suitability

Mother with infant, Tuanan, Central Kalimantan. **Photo** Perry van Duijnhoven

Flanged male, Tuanan, Central Kalimantan.
Photo Perry van Duijnhoven

Aerial image of a forest and oil-palm plantation in Sabah, Malaysia.

Photo Conservation Drones

CLIMATE CHANGE IMPACTS ON THE SUITABILITY OF LAND FOR OIL PALM

Oil palm and orangutan habitat

Conversion of forests, whether primary or degraded, to oil-palm plantations is a major threat to remaining orangutans (WICH *et al.* 2012) and other biodiversity (MEIJAARD & SHEIL 2013; FOSTER *et al.* 2011; FITZHERBERT *et al.* 2008). Historically, the main growing regions for oil palm on Borneo have been in West Kalimantan and Sabah, but increasing demand has spurred a drive for planting in other regions, primarily in Sarawak, as well as Central and East Kalimantan. Demand for oil-palm products continues to increase and is projected to increase further in the future (CORLEY 2009). Therefore, long-term conservation planning for Borneo's biodiversity should account for potential changes to the extent of this threat as well as the responses of species to the resulting environmental change (KOH & SODHI 2010).

Oil palm grows in a relatively narrow bioclimatic zone in the tropics that in Southeast Asia largely overlaps with suitable habitat for orangutans (FITZHERBERT *et al.* 2008). As such, orangutans face competition for land from oil palm producers. While economic, legal and environmental considerations have a large role in deciding where oil palm can and cannot be planted (SMIT *et al.* 2013), factors concerning crop productivity are also important (CORLEY & TINKER 2003). Crop productivity, and importantly, yield, is a function of soil quality, topography and climate. Of these attributes, those that are related to climate are expected to change over time, yet recent efforts to map oil-palm suitability (*e.g.* ARSHAD *et al.* 2012; GINGOLD *et al.* 2012; MANTEL *et al.* 2007) have kept these attributes fixed to current values.

Of all climatic attributes, those related to rainfall have the largest influence on oil-palm productivity (CORLEY & TINKER 2003). The length of the dry season has historically been a limiting factor in site selection. Seasonal droughts are linked with reduced yields as water-stressed palms tend to produce fewer female flowers, and unripe fruits are aborted (BASIRON 2007). During the early phases of plantation establishment, water stress also inhibits the growth of oil-palm saplings (CAO *et al.* 2011). At the other extreme, too much rainfall results in sporadic flooding, and leads to soil erosion, which has knock-on impacts on yield (CORLEY & TINKER 2003). The suitability of land for oil-palm cultivation is likely to change as the climate alters. While changes in temperature will not particularly affect the oil palm, the alterations in rainfall patterns predicted by climate models will in part determine whether land is appropriate for establishing this crop. Because oil-palm expansion is a threat to orangutan populations, it is important to anticipate whether land deemed suitable or unsuitable for oil palm in the future overlaps with suitable orangutan habitat.

The World Resources Institute (WRI) and Sekala recently used a simple approach to create an online mapping tool to identify land suitable for oil-palm cultivation (GINGOLD *et al.* 2012). For land to be considered suitable for growing oil palm certain environmental, legal, and socio-economic factors were taken into consideration. The main delimiting criteria included factors concerning crop productivity, defined by elevation and slope; soil depth, soil type, and acidity; as well as mean annual rainfall. They also took into account provisions for biodiversity conservation and existing carbon storage in forests.

The potential limitations on oil-palm production posed by changes to rainfall patterns are examined here by extending the WRI/Sekala approach to include future rather than current rainfall patterns. Rainfall data from the CSIRO model is incorporated under a best-case emission scenario. These data exhibit the least change over time amongst the various models examined.

Oil-palm suitability under changing climatic conditions

According to the assessment criteria, a reduction in the amount of productive land for oil palm is to be expected on Borneo due to elevated rainfall over the course of the century. Much of the area deemed suboptimal for the crop is also expected to become more seasonal according to climate projections, which suggests that elevated rainfall, though modest in areas, could be accompanied by greater risk of flooding and drought, which would also impact oil-palm yields.

Suboptimal growing regions currently include the majority of Sarawak (Figure 18). This is in part the result of high rainfall in this part of Borneo, but also due to the large quantity of peat lands in coastal areas, which have a low agricultural value, a high carbon value, and are prone to flooding. A projected rise in annual rainfall in this growing region would further reduce oil-palm productivity. It is therefore notable that, despite its suboptimal growing potential, clearance of forest for plantations in coastal Sarawak is extensive and ongoing (KOH *et al.* 2011).

The increased rainfall expected under climate projections could also result in currently productive growing regions being substantially reduced, particularly in the interior parts of Central and West Kalimantan (Figure 18). The main districts that would be affected in this way are Barito Selatan, Barito Utara and Gunung Mas in Central Kalimantan, and Sintang, Sekadau and Kapuas Hulu districts in West Kalimantan. Higher annual rainfall, in excess of 3500 mm/year, is projected in these regions by 2050, with little additional change to the area affected by 2080. Large parts of the core oil-palm growing regions on Borneo are expected to remain optimal for cultivation – principally in the West, South and East Kalimantan lowlands, Sabah and to some extent parts of Central Kalimantan. Much of this land, particularly in West Kalimantan and Sabah, is already planted, suggesting that the optimal areas for oil-palm production have long been recognized.

It is important to note that these predictions of future oil-palm productivity and suitability are based on present-day growth characteristics of the various commercially available oil-palm cultivars, and do not take into account potential genetic improvement of plants that could increase drought or flood resistance. Such developments are ongoing in the industry and vital to the long-term sustainability of the oil-palm crop (COCHARD *et al.* 2005). While drought or flood resistant cultivars will likely be developed in due course, productivity assessments such as the one presented here may still be informative for parts of the industry in which uptake of new cultivars might be slow, such as for small-holders.

Spatial assessments of agricultural suitability are also useful to identify where land-use conflicts are likely to occur in orangutan habitat on Borneo. They are also informative for identifying likely strongholds for the species in an era of environmental change.

These data provides a mixed outlook for the species. On one hand a large part of the orangutan range will continue to be comprised of prime land for growing oil palm, particularly in the interiors of all range states (Figure 18B). On the other hand, core components of the orangutan range should continue to be highly suitable for the species while providing suboptimal conditions for oil-palm cultivation. Unfortunately, with the exception of the Kapuas lakes and foothills of the Madai plateau in West Kalimantan, these areas are fragmented across the orangutan range.

Figure 18 **Change in suitability of land for growing oil palm on Borneo due to projected changes in rainfall.**
Maps on the left pane (A) show suitable land for current and future time periods according to crop productivity criteria (elevation, slope, soil characteristics and annual precipitation) as well as additional environmental constraints arising from carbon, watershed protection and biodiversity provisions. The former criteria are indicated as 'productive' land, and the latter are indicated as 'suitable'. Rainfall is the only limiting factor that changes in future time periods due to projected climate changes (CSIRO model under a best-case scenario). The right pane (B) shows the productivity criteria overlaid with the suitable habitat coverage for the orangutan. For clarity, only areas within the known current core range are presented.

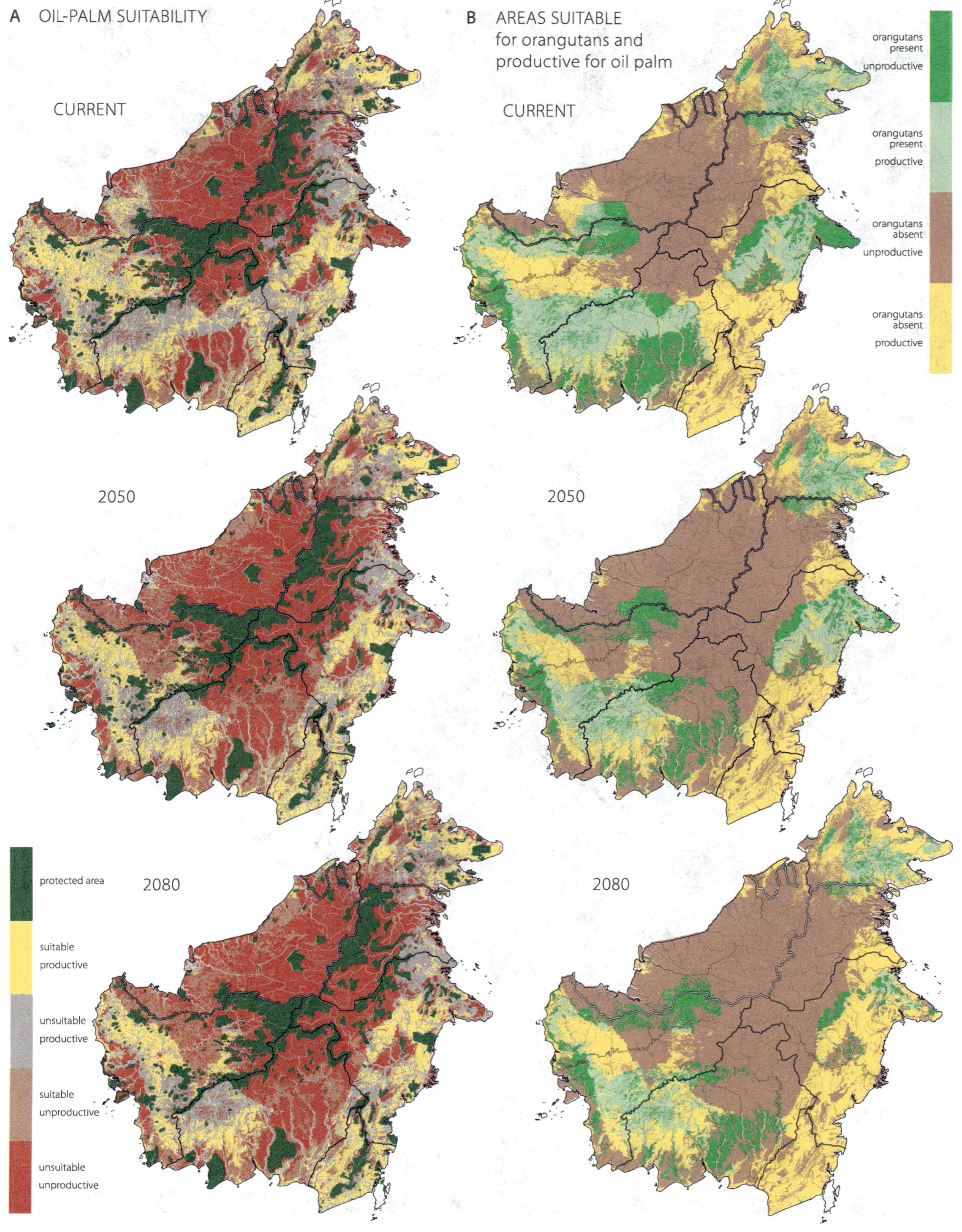

A OIL-PALM SUITABILITY
CURRENT
2050
2080
protected area
suitable productive
unsuitable productive
suitable unproductive
unsuitable unproductive
B AREAS SUITABLE for orangutans and productive for oil palm
CURRENT
2050
2080
orangutans present unproductive
orangutans present productive
orangutans absent unproductive
orangutans absent productive

Unflanged male, Tuanan, Central Kalimantan.

DISCUSSION

Despite five decades of conservation attention focused on the orangutan, their numbers are still declining. Even though population data barely existed 50 years ago, it is widely accepted that every year, large areas of orangutan habitat are degraded or lost on both Sumatra and Borneo (GAVEAU *et al.* 2014; WICH *et al.* 2011; MEIJAARD & Wich 2007) and that up to 3 000 orangutans on Borneo are killed annually (DAVIS *et al.* 2013; MEIJAARD *et al.* 2011). By 'hindcasting' habitat suitability predictions prior to the 1950s when much of Borneo's forests were intact, models estimate that up to 316 000 km² of land was suitable for orangutans within the current known core range. As of 2010 deforestation led to approximately 18% of this habitat being lost, leaving the current extent of suitable habitat at 260 000 km². Further deforestation predicted for this century could lead to a further loss of 15.5% by 2080, leaving up to 219 000 km² of habitat remaining within the current core orangutan range. It is important to note that these estimates of suitable habitat are nevertheless conservative, and they also include degraded areas where forest cover is patchy. It is likely that orangutans remaining in these areas would subsist at low densities.

In addition to land-cover change and hunting, there is another significant threat that will add to the pressures facing orangutans in the coming decades: Spatial models point to the possibility that a large amount of current orangutan habitat will become unsuitable because of changes in climate (STRUEBIG *et al.* 2015). The extent of these changes depends on various climate projections, yet even the best-case emission scenario we assessed leads to losses of suitable habitat for orangutans under both climate models. The amount of suitable habitat available for Borneo's orangutans in the future is likely to be much lower when climate projections are considered compared to the losses from the effects of deforestation alone (Figure 19). Across all climate and land cover change projections we assessed, models predict that between 49 000 and 83 000 km² of orangutan habitat might remain by 2080, reflecting a loss of around 69-81% since 2010. This projection represents a three to five-fold greater decline in habitat than that predicted from deforestation projections alone.

Yet, despite these impacts of predicted land-cover and climate change on orangutans there is much that can be done to prevent the extinction of populations. Although climate change impacts are the most difficult to prevent, even if global action on reducing emissions of greenhouse gases is taken. Orangutans are, however, ecologically versatile and the species may to a certain extent be able to cope with ecological change induced by climate change alone. How this would affect population dynamics and survival is difficult to predict, but it is obvious that reducing deforestation would significantly increase the likelihood of population survival. The safeguarding of remaining wild orangutan populations on Borneo depends on the development and implementation of more appropriate land-use policies that stop the destruction and degradation of orangutan habitats (MEIJAARD *et al.* 2012).

Prioritising habitat for conservation

The coastal lowland parts of West, Central and East Kalimantan provide key habitat for orangutans, and yet are projected to experience the brunt of Borneo's forthcoming deforestation by 2080. In these areas several large populations persist, most notably the population in the Sabangau-Katingan peatland (Figure 20), which supports an estimated 9 900 individuals (WICH *et al.* 2008). Much of the forest across this region is legally gazetted for conversion purposes, and areas outside of this designation have already experienced some deforestation in the past despite being afforded some form of protection (GAVEAU *et al.* 2013). Of all the orangutan populations on Borneo it is arguably this population that requires most immediate attention as a major stronghold of the species.

Over time, a major reduction in the extent of suitable orangutan habitat can be expected. Yet there are core strongholds of suitable orangutan habitat that are predicted to remain to the west, east and northeast of the island where populations of *P. p. wurmbii* and *P. p. morio* are found (Figure 20). Additional areas of Borneo where orangutans are not currently thought to live are likely to remain theoretically suitable habitat in the future (Figure 17), although it is unclear why no populations are found there at the moment.

The dramatic projections presented in this study point to the necessity of immediate and substantial action to protect the Bornean orangutan. In order to succeed new, pragmatic ways to manage orangutan populations in the wild may need to be explored. One such option for orangutan conservation is to explore the possibility of expanding populations into the climatically suitable areas which are not currently inhabited by the species. This 'assisted colonisation' is not an easy task and should be considered a last resort. Successful colonisation would require highly detailed feasibility assessments for each potential introduction region, to include studies of forest phenology, ecological impact, and threats, including hunting (BECK *et al.* 2007).

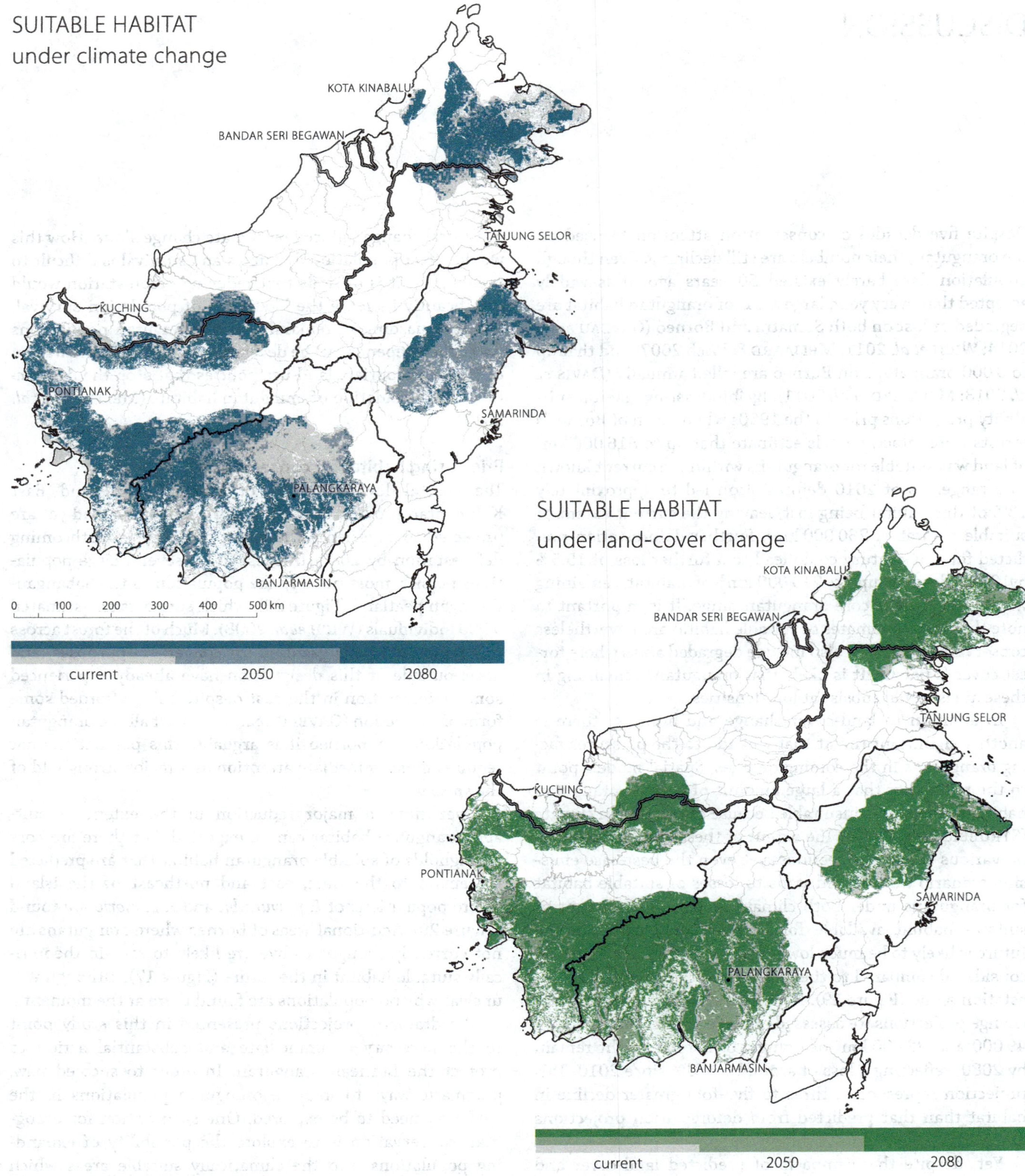

Figure 19 **Projected decline in suitable orangutan habitat forecasts of land cover and/or climate change.**
Each map presents the predicted change in suitable habitat within the known core range of orangutan this century. The darkest shading indicates the land predicted to remain suitable by 2080, with lighter shades indicating change from suitable to unsuitable.

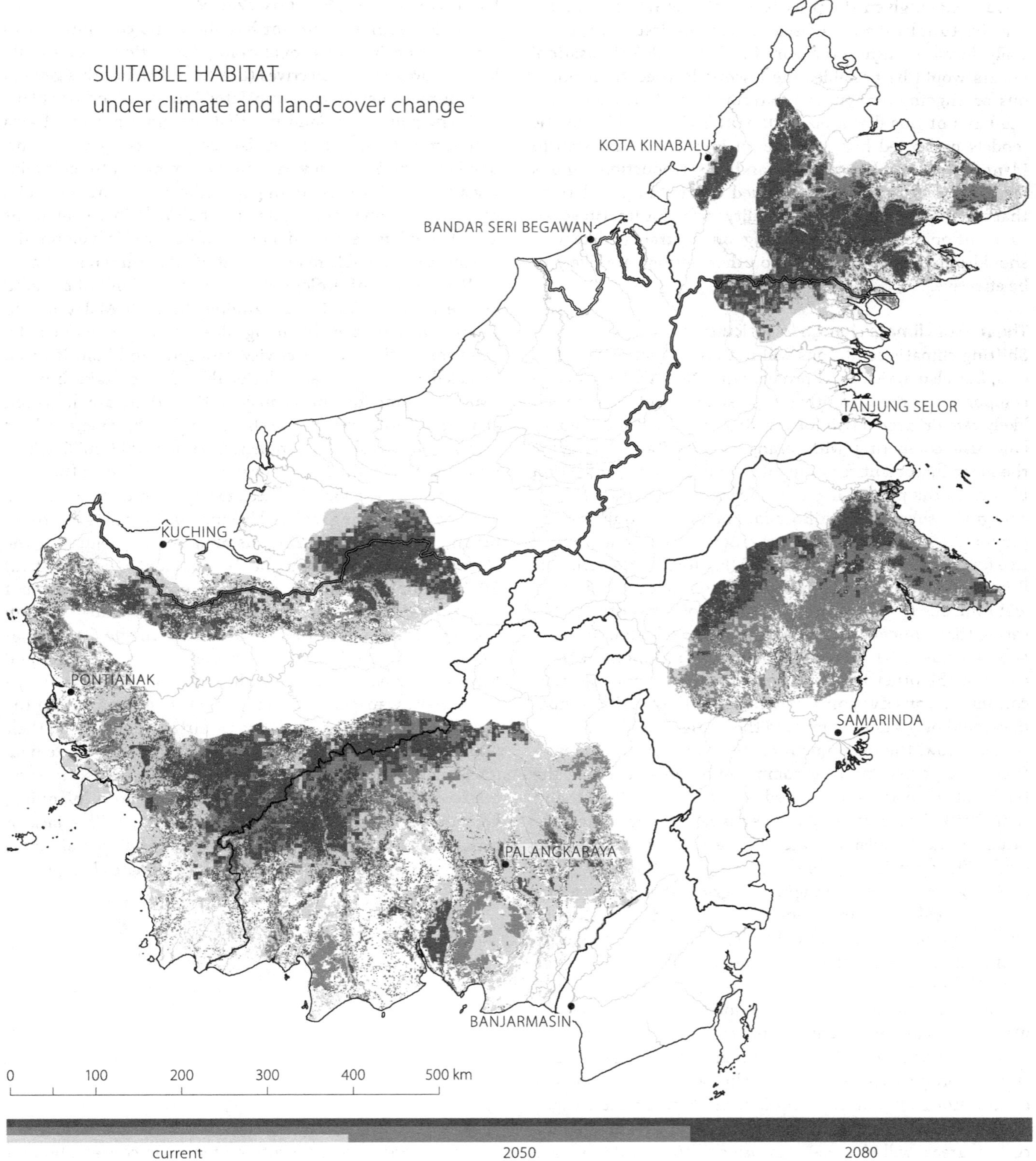

SUITABLE HABITAT
under climate and land-cover change
KOTA KINABALU
BANDAR SERI BEGAWAN
TANJUNG SELOR
KUCHING
PONTIANAK
SAMARINDA
PALANGKARAYA
BANJARMASIN
0   100   200   300   400   500 km
current
2050
2080

Nevertheless, given the predictions of this report it seems sensible to at least start a discussion on this issue and potentially develop a legal framework in which such 'colonisation' efforts would be possible. The lessons learned from previous or ongoing orangutan reintroduction efforts can form the basis of this discussion (RUSSON 2009). The bioclimatic models presented here provide an important first step in identifying possible expansion and/or introduction regions for each of the island's threatened orangutan populations that will maintain climatic suitability in the future. However, as mentioned before, embarking on a strategy like this should only be considered if no other options are likely to be effective.

The role of climate change in agriculture

Shifting climatic conditions will not only affect natural forests, but also agricultural productivity. Nonetheless, a large component of the currently recognised orangutan range will likely remain productive for growing oil palm – the primary land use competing with orangutan habitat. Overlaying the areas important for orangutan and oil palm production illustrates the potential extent of this conflict for the three orangutan subspecies on Borneo. Fortunately, only a small proportion of the *P. p. pygmaeus* range will likely be productive for oil-palm cultivation in 2080 (Figure 21). However, for *P. p. morio* and *P. p. wurmbii* large parts of their current range will continue to be suitable at the close of the century, indicating that competition for land from the oil-palm industry will continue to be a conservation problem for these subspecies. On the other hand, this overlap also emphasises that continuing conservation partnerships with oil palm companies could play a defining role in the orangutan's future.

Even now, the main environmental impact of deforestation, as reported by local communities on Borneo, is the temperature increase associated with forest loss (MEIJAARD *et al.* 2013). These are local temperature effects not incorporated in the broader climate change models used in this study. They are highly relevant though at the local level, affecting issues as wide-ranging as agricultural outputs, disease outbreaks, fire and smoke impacts, and water supply. There is a growing need for political planning and regulation to start incorporating these local and regional climate change impacts, and mitigate or adapt through changed land-use planning and implementation. Ignoring these factors will ultimately cause major environmental, economic, and social impacts. This planning includes anticipating changes in agricultural productivity in the oil-palm sector, an important contributor to Indonesia's and Malaysia's economic output. Reduced yields as a result of climate change, as predicted for certain areas, will increase pressure on other lands that are climatically more suitable. This chain of events could lead to further deforestation and thus exacerbate negative environmental impacts. Better planning for oil palm is therefore needed that incorporates land suitability (including future climatic conditions), as well as social and environmental impacts of oil palm (SMIT *et al.* 2013).

Land-use planning for conservation

The Indonesian government has shown its commitment to conservation in its national orangutan action plan and its low carbon growth objectives (ELSON 2011), but this general commitment needs to be translated into new land-use plans and new policies on land use that integrate protected area management with broader landscape-level management. Similarly, the Sabah government is committed to maintaining all remaining orangutan populations, as expressed in its state action plan for orangutans (Sabah Wildlife Department 2011), and has developed a range of policy briefs emphasising more sustainable management of its natural resources.

The concept of ecological connectivity should be placed at the heart of land-use planning. In both Malaysia and Indonesia land-use planning does not yet sufficiently incorporate these connectivity concepts, and land is often parceled out in discrete blocks that do not take broader landscape functions into consideration. There are, however, signs that this trend may be changing. Sabah, for example, is creating the Sabah Development Corridor 2008-2025, which was initiated by the State Government to enhance the quality of life for the people by accelerating the growth of Sabah's economy, promoting regional balance and bridging the rural-urban divide while ensuring sustainable management of the state's resources (Institute for Development Studies (Sabah) 2007). Implementation of this resource management strategy can be seen in such projects as the 'Kinabatangan Corridor of Life' and the 'Ecolink Assessment Study' that evaluate possible linkages between Crocker Range and Kinabalu Parks. In Indonesia, the development of landscape level forest management units (KPH) is another step in the right direction. So far, these are all early initiatives and their final implementation and potential effectiveness remain unclear. Furthermore, unless such developments are integrated with others, such as road infrastructure planning, landscape-level connectivity may still decline. This lack of connectivity would leave wildlife with fewer dispersal routes and strongly reduced means to adapt to climate change.

Figure 20 **Focal areas for future orangutan conservation on Borneo based on climate and land-cover change projections.** Many areas are under substantial pressure for conversion to agriculture, and it is important to note that some will remain productive for oil-palm cultivation in years to come. The Sabangau-Kahayan peatswamps (brown) support one of the largest known orangutan populations on Borneo, but the area is vulnerable to land cover and

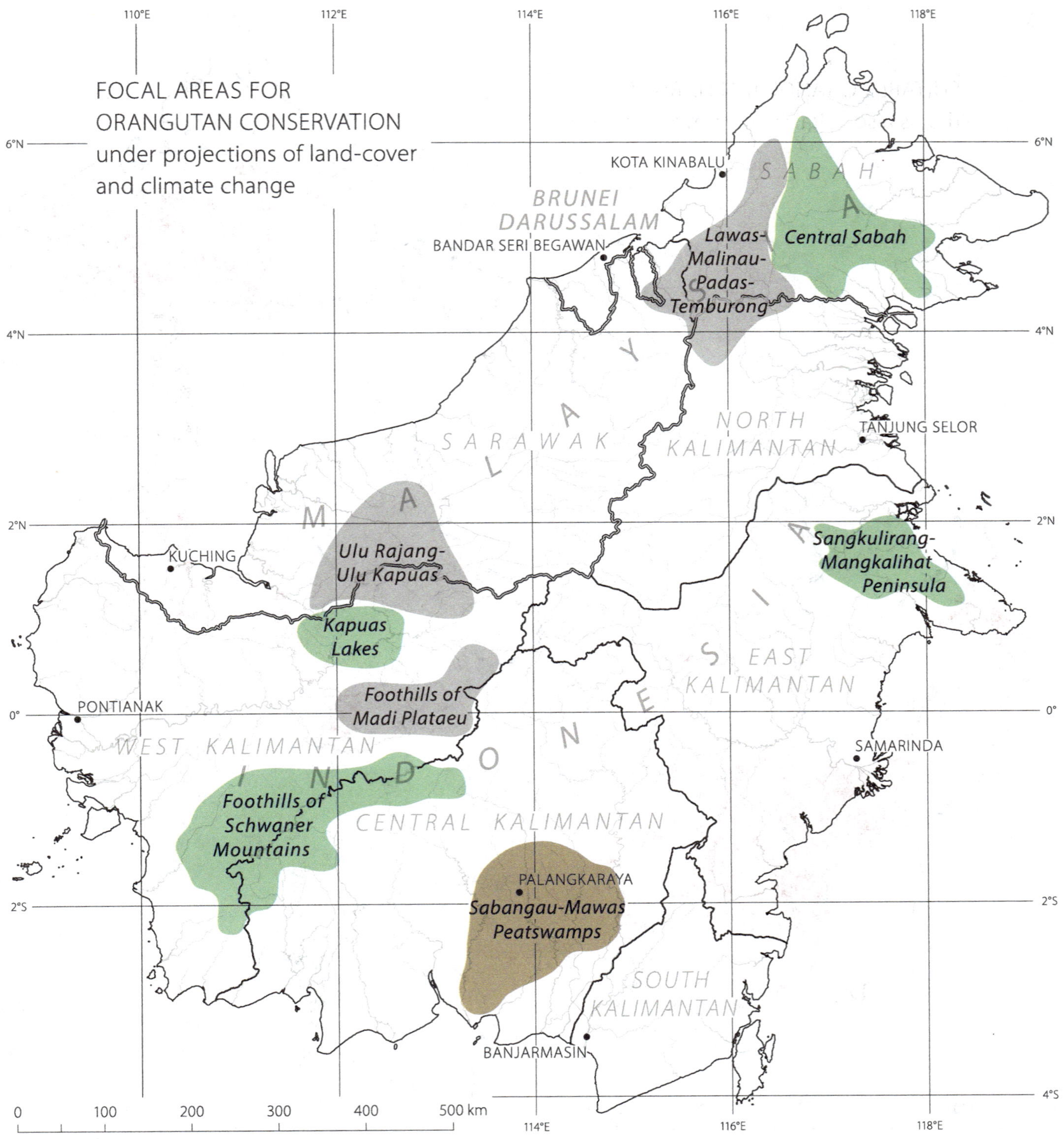

climate changes projected for this century. For each sub-species there are parts of Borneo (green) that should remain as strongholds for orangutans during the environmental changes predicted. These areas are likely to become more suitable over time, especially those at mid-elevation. Areas indicated in grey currently do not support orangutans but, according to models presented here, habitat might be suitable and continue to be in the future despite projected climate changes. Notably, projected increases in rainfall might place greater constraints on the productivity of land for oil palm cultivation in all regions except for western Sabah and the Sangkulirang-Mangkalihat peninsula in East Kalimantan. Even though the upper Kapuas Lakes were identified as areas with relatively limited predicted climatic change, recent conversion for oil palm around the lakes area has resulted in major losses of orangutan habitat.

53

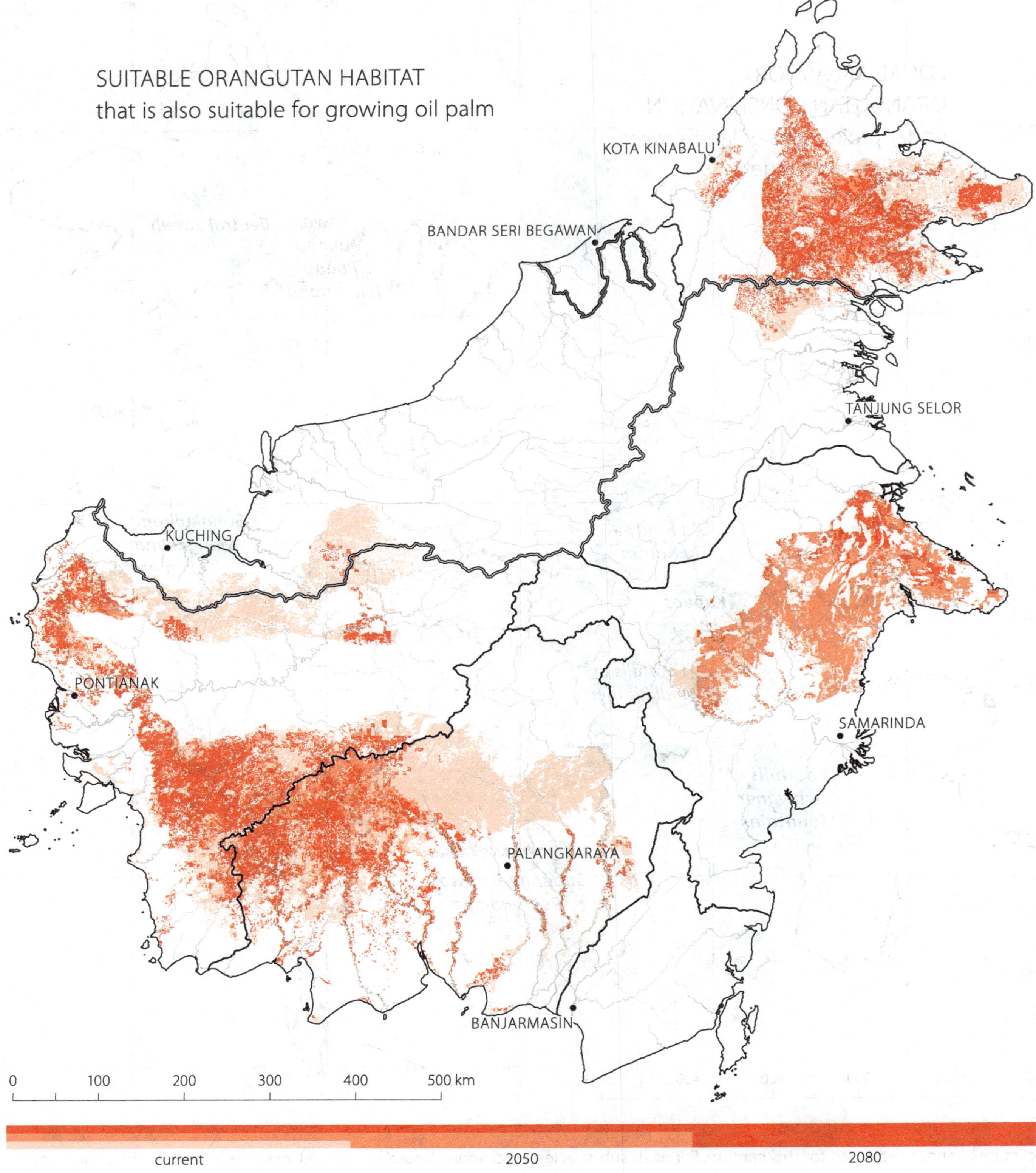

Figure 21   **Overlays of the land predicted to be suitable for orangutan under land-cover and climate change models, but also productive for oil palm.**
Climate conditions used are for the CSIRO projections under the best-case emission scenario.

Law enforcement, coupled with forest management that ensures that plans are indeed implemented according to government regulations is an essential accompaniment to any land-use planning that considers the broader ecological functions of forests and biodiversity. The Malaysian and Indonesian governments' law enforcement and protected area management are insufficient at the moment. Governments need to seriously consider how they can improve their effectiveness in both of these areas. Ensuring that those that break the law are actually caught and prosecuted is vital and would require 1) training of police, judges, and conservation authorities; 2) targeting corruption in government offices; and 3) ensuring that the public understand why laws are taken seriously now when they were not previously. A recent study suggested that 27% of the people in Kalimantan did not know that orangutans are protected by law (MEIJAARD et al. 2011) therefore campaigns to effectively inform the public are also vital.

Law enforcement measures are particularly important in protecting wildlife within timber concessions in natural forests. Through sustainable management of these areas, large areas of productive forest can be maintained as well as the environmental services they provide (WILSON et al. 2010). However, for orangutans to benefit from the maintenance of logged forests, these do need to be sustainably managed, which many are not. Governments would need to increase the auditing of timber-concession management, punish those that break regulations and manage their forests unsustainably, and reward those who manage their concessions well with financial or regulatory incentives. In addition, much needs to be improved in the management of protected areas (GAVEAU et al. 2013), which would ensure that at least these areas provide genuine safe zones for orangutans and other threatened wildlife.

Finally, laws relating to land-use planning need to become more flexible to be able to adapt land use to a changing climate (LACKMAN and MURPHY 2014). Currently, allocation of land to either permanent forest (for protection or production) or non-forest is based on a range of laws like the Sabah Forest Enactment from 1968 or Indonesia's basic Forest Law of 1980. These laws use various criteria to determine which lands should remain permanently forested and which can be converted to non-forest uses. For example, Indonesia's law for determining which forests should be protected for maintaining watershed services uses rain intensity as its defining criterion in addition to soil type and slope. Shifting patterns of rainfall and rainfall intensity, as indicated in this report, would therefore require areas to undergo constant adaptations in land-use allocations. Such a flexible approach to land-use planning under changing climatic conditions does not yet exist in either Malaysian or Indonesian Borneo. In addition, once an area has been licensed for forest use or conversion rather than strict protection there are significant legal constraints and associated major costs to reverse such licenses. Predictive modeling in conjunction with early policy adaptations are thus needed to minimize the impacts that climate change could have on wildlife in Borneo.

People and the orangutan

Some of the most important stakeholders in orangutan conservation are the estimated 5 million people in Borneo that live alongside the orangutans. They are the least engaged sector. Borneo's people still largely depend to a large degree on forest resources, for their livelihoods, health and spiritual well-being (MEIJAARD et al. 2013). The aspirations and day-to-day needs of forest-based communities will play a central role in the long-term maintenance of Borneo's forests and wildlife. Outside of protected areas, the survival of the orangutan depends on their peaceful coexistence with these communities, particularly in consideration of the pressures caused by hunting (CAMPBELL-SMITH et al. 2011; CAMPBELL-SMITH et al. 2010). If current rates of orangutan killings continue, many populations could go extinct within one human generation, regardless of any habitat management interventions. Unless orangutan killing is reduced, only those few populations that are well-protected or that occur in areas where killings are rare due to cultural and traditional reasons in the local community (e.g. Kinabatangan or other populations located in eastern Sabah) stand a chance of long-term survival (ANCRENAZ et al. 2007; ANCRENAZ et al. 2004). Encouraging rural people to support the principles of environmental conservation and be actively responsible for the management of their resources is therefore a crucial requirement for successful orangutan conservation (ANCRENAZ et al. 2007).

Respecting and formalizing people's rights to forest use would be a first step towards stabilizing Borneo's rapidly disappearing forest frontiers, while laying a basis for two-way discourse about how the challenge of orangutan killing could best be addressed (MEIJAARD et al. 2013).

Paying for conservation

All of this indicates that governments need to develop much more holistic policies that not only target economic progress, but balance that progress with ecological and environmental sustainability. In the case of Borneo, for example, it could strive to prevent an increase of flooding which is now leading to high humanitarian and economic costs (WELLS et al. 2013). Such an approach could also aim to put a monetary value on forests. Until markets pay for environmental services provided by forests, exploiting these forests or replacing them with more productive uses is always going to generate more revenue than conserving them. Some of these opportunity costs can be offset by legislation or ethics – i.e. "we are legally obliged to protect these forests", or "we feel it is the right thing to do" – but it is likely that in most situations the protection or sustainable management of orangutan habitats will have to be paid for. Carbon markets may provide some of the financial means required for this payment, while other sources of income could come from tourism, payments for water, and other forest related economic activities (WICH et al. 2011, VENTER et al. 2009). Companies could also be legally required to fund a significant part of the total cost by implementing orangutan-friendly management on their land (DENNIS et al. 2010).

Riparian vegetation set aside conservation areas in new oil-palm plantations, Sabah. **Photo** Matthew Struebig

The role of hunting and some caveats

The orangutan habitat modelling in this study is based on the assumption that orangutan distribution is significantly correlated with environmental and climatological factors. Even though the model fit is generally high, indicating that these environmental and climatological factors play an important role in determining where orangutans occur, there are additional factors that influence their distribution including the killing of orangutans. Recent studies have highlighted that deliberate killing of orangutans is a common occurrence throughout much of Borneo: up to 3 000 orangutans are estimated to be killed each year (MEIJAARD *et al.* 2011). There is a variety of cultural and/or economic reasons why humans kill orangutans, and certain areas have much greater incidences of killing than others (DAVIS *et al.* 2013; MEIJAARD *et al.* 2011). The extent to which present and past killing have shaped the distribution range of orangutans remains unclear, although it has been speculated that historic hunting has played a major role in the population declines, and sometimes local extinction, of many populations (MEIJAARD *et al.* 2010). Overall orangutan mortality rates in Kalimantan seem to significantly exceed the maximum rates that populations of this slow-breeding species can sustain (MEIJAARD *et al.* 2011; MARSHALL *et al.* 2009). Therefore all hunted populations are expected to decline, irrespective of what happens to their habitat due to land cover or climate change, if hunting pressure does not decline. The trends presented in this report, based on land cover and climate change projections, must therefore be evaluated in the light of killing patterns. In particular in many areas in Borneo's uplands identified as optimal for conservation under land-cover and climate change projections, the probability of orangutan killings is also high. If for ecological reasons orangutans are increasingly restricted to these areas, there would be even more urgent need to ensure that human-induced mortality rates in orangutans are minimized. DAVIS *et al.* (2013) provide useful recommendations as to what strategies can be locally employed depending on the main reasons for why killings occurs (*e.g.* for food, or because of agricultural conflicts).

Rehabilitating youngster, Nyaru Menteng, Central Kalimantan.

Photo Perry van Duijnhoven

RECOMMENDATIONS

1   Identify all remaining orangutan populations and undertake an assessment of their viability, based on local threat levels from killing and incompatible land use, present and predicted habitat quality, and population dynamics (dispersal, inbreeding, etc.). Identify all orangutan populations that the governments officially agree to maintain.

2   Conduct a triage process in which the 'must save' and 'can save' populations are identified and agreed by governments, both inside and outside protected areas, as well as the management measures needed to maintain them.

3   Based on the above, propose and designate new protected areas or other areas under permanent forest cover that are large and safe enough to contain viable orangutan populations.

4   Where possible, connect these permanently forested areas through uninterrupted forested corridors, for example in permanent natural forest timber concessions, that allow orangutans and other wildlife to move through the landscape in reaction to changing climatic and ecological conditions.

5   Reconcile these land-use plans with other spatial plans (for development, infrastructure, agriculture, etc.), and endorse these planned land uses in high level government regulations that allocate special strategic status to these orangutan populations, and which are strong enough not to be overruled by other national-level or local level regulations.

6   Effectively enforce laws regarding the killing of orangutans and implement public campaigns and other communication efforts that make the public aware of the illegality of killing.

7   Seek innovative ways to augment protected areas to conserve remaining orangutan forests, which could include, for example, leverage of carbon mitigation investments via the UN-REDD+ (Reducing Emissions from Deforestation and Forest Degradation) mechanism, or other payments for forest environmental services (e.g., erosion control, flood buffering, local climate regulation). Such systems might not only reduce threats to biodiversity from land-cover changes but also contribute to local community welfare and safety and to international efforts to reduce climate change.

8   Drained coastal peatlands in Borneo are predicted to decompose and flood leaving behind unproductive brackish swamps. Peatlands are key orangutan habitats that should be left forested and undrained to avoid major negative biodiversity and socio-economic impacts.

REFERENCES

ACHARD, F., H.D. EVA, H.-J. STIBIG, P. MAYAUX, J. GALLEGO, T. RICHARDS and J.-P. MALINGREAU. 2002. *Determination of deforestation rates of the world's humid tropical forests.* Science **297**:999-1002.

ALI, R. and S.M. JACOBS. 2007. *Saving the rainforest through health care: Medicine as conservation in Borneo.* International Journal of Occupational & Environmental Health **13**:295-311.

ANCRENAZ, M., F. ORAM, L. AMBU, I. LACKMAN, E. AHMAD, H. ELAHAN, and E. MEIJAARD. 2014. *Of pongo, palms, and perceptions – A multidisciplinary assessment of Bornean orang-utans in an oil palm context.* Oryx. DOI: http://dx.doi.org/10.1017/S0030605313001270.

ANCRENAZ, M., L. AMBU, I. SUNJOTO, E. AHMAD, K. MANOKARAN, E. MEIJAARD and I. LACKMAN. 2010. *Recent surveys in the forests of Ulu Segama Malua, Sabah, Malaysia, show that orang-utans (P. p. morio) can be maintained in slightly logged forests.* PLoS ONE **5**:e11510.

ANCRENAZ, M., L. DABEK and S. O'NEIL. 2007. *The costs of exclusion: Recognizing a role for local communities in biodiversity conservation.* PLoS Biology **5**:2443-2448.

ANCRENAZ, M., O. GIMENEZ, L. AMBU, K. ANCRENAZ, P. ANDAU, B. GOOSSENS, J. PAYNE, A. SAWANG, A. TUUGA and I. LACKMAN-ANCRENAZ. 2004. *Aerial surveys give new estimates for orangutans in Sabah, Malaysia.* PLoS Biology **3**:e3.

ANSHARI, G., A.P. KERSHAW and S. VAN DER KAARS. 2001. *A late Pleistocene and Holocene pollen and charcoal record from peat swamp forest, Lake Sentarum Wildlife Reserve, West Kalimantan, Indonesia.* Palaeogeography, Palaeoclimatology, Palaeoecology **171**:213-228.

ARSHAD, A.M., M.E. ARMANTO and A.M. ZAIN. 2012. *Evaluation of climate suitability for oil palm (Elaeis guineensis Jacq.) cultivation.* Journal of Environmental Science and Engineering **1**:272-276.

BAPPENAS 2010. *Indonesia climate change sectoral roadmap (ICCSR).* Badan Perencanaan Pembangunan Nasional, Indonesia, Jakarta, Indonesia.

BARNOSKY, A.D., N. MATZKE, S. TOMIYA, G.O. WOGAN, B. SWARTZ, T.B. QUENTAL, C. MARSHALL, J.L. MCGUIRE, E.L. LINDSEY and K.C. MAGUIRE. 2011. *Has the Earth's sixth mass extinction already arrived?* Nature **471**:51-57.

BASIRON, Y. 2007. *Palm oil production through sustainable plantations.* European Journal of Lipid Science and Technology **109**:289-295.

BECK, B.B., K. WALKUP, M. RODRIGUES, S. UNWIN, D. TRAVIS and T. STOINSKI. 2007. *Best practice guidelines for the reintroduction of great apes.* IUCN.

BELLARD, C., C. BERTELSMEIER, P. LEADLEY, W. THUILLER and F. COURCHAMP. 2012. *Impacts of climate change on the future of biodiversity.* Ecology letters **15**:365-377.

BROOK, B.W., C.J.A. BRADSHAW, L.P. KOH and N.S. SODHI. 2006. *Momentum drives the crash: Mass extinction in the tropics.* Biotropica **38**:302-305.

BROOK, B.W., N.S. SODHI and C.J.A. BRADSHAW. 2008. *Synergies among extinction drivers under global change.* Trends in Ecology and Evolution **23**:453-460.

CAMPBELL-SMITH, G., M. CAMPBELL-SMITH, I. SINGLETON and M. LINKIE. 2011. *Apes in space: Saving an imperilled orangutan population in Sumatra.* PLoS ONE **6**:e17210.

CAMPBELL-SMITH, G., H.V.P. SIMANJORANG, N. LEADER-WILLIAMS and M. LINKIE. 2010. *Local attitudes and perceptions toward crop-raiding by orangutans (Pongo abelii) and other nonhuman primates in Northern Sumatra, Indonesia.* American Journal of Primatology **72**:866-876.

CAO, H.-X., C.-X. SUN, H.-B. SHAO and X.-T. LEI. 2011. *Effects of low temperature and drought on the physiological and growth changes in oil palm seedlings.* African Journal of Biotechnology **10**:2630-2637.

CARATINI, C. and C. TISSOT. 1988. *Paleogeographical evolution of the Mahakam delta in Kalimantan, Indonesia during the Quaternary and late Pliocene.* Review of Palaeobotany and Palynology **55**:217-228.

CARLSON, K.M., L.M. CURRAN, G.P. ASNER, A.M. PITTMAN, S.N. TRIGG and J. MARION ADENEY. 2013. *Carbon emissions from forest conversion by Kalimantan oil palm plantations.* Nature Clim. Change **3**:283-287.

CHEN, I.-C., J.K. HILL, R. OHLEMÜLLER, D.B. ROY, and C.D. THOMAS. 2011. *Rapid range shifts of species associated with high levels of climate warming.* Science **333**:1024-1026.

CHEN, I.-C., H.-J. SHIU, S. BENEDICK, J.D. HOLLOWAY, V.K. CHEY, H.S. BARLOW, J.E. HILL and C.D. THOMAS. 2009. *Elevation increases in moth assemblages over 42 years on a tropical mountain.* Proceedings of the National Academy of Sciences of the United States of America **106**:1479-1483.

COCHARD, B., P. AMBLARD and T. DURAND-GASSELIN. 2005. *Oil palm genetic improvement and sustainable development.* Oléagineux, Corps Gras, Lipides **12**:141-147.

COLWELL, R.K., G. BREHM, C.L. CARDELIUS, A.C. GILMAN and J.Y. LONGINO. 2008. *Global warming, elevational range shifts, and lowland biotic attrition in the wet tropics.* Science **322**:258-261.

CORLEY, R. 2009. *How much palm oil do we need?* Environmental Science & Policy **12**:134-139.

CORLEY, R.H.V. and P.B. TINKER. 2003. *The oil palm.* Blackwell Science.

DAVIS, J.T., K. MENGERSEN, N. ABRAM, M. ANCRENAZ, J. WELLS and E. MEIJAARD. 2013. *It's not just conflict that motivates killing of orangutans.* PLoS ONE **8**:e75373.

DENNIS, R., A. GRANT, Y. HADIPRAKARSA, P. HARTMAN, D.J. KITCHENER, T. LAMROCK, F. MACDONALD, E. MEIJAARD and D. PRASETYO. 2010. *Best practices for orangutan conservation – Oil palm concessions.* Orangutan Conservation Services Program, Jakarta, Indonesia.

DEPARTEMEN KEHUTANAN. 2007. *Strategi dan rencana aksi konservasi orangutan Indonesia 2007-2017.* Direktorat Jenderal Perlindungan Hutan Dan Konservasi Alam, Departemen Kehutanan (Indonesian Ministry of Forestry), Jakarta, Indonesia.

ELSON, D. 2011. *Cost-benefit analysis of a shift to a low carbon economy in the land-use sector in Indonesia.* UK Climate Change Unit of the British Embassy, Jakarta, Indonesia.

FITZHERBERT, E.B., M. STRUEBIG, A. MOREL, F. DANIELSEN, C. BRÜHL, P.F. DONALD and B. PHALAN. 2008. *How will oil palm expansion affect biodiversity?* Trends in Ecology & Evolution **23**:538-545.

FLENLEY, J. 1998. *Tropical forests under the climates of the last 30,000 years.* Climatic Change **39**:177-197.

FOSTER, W.A., J.L. SNADDON, E.C. TURNER, T.M. FAYLE, T.D. COCKERILL, M.D. F. ELLWOOD, G.R. BROAD, A.Y.C. CHUNG, P. EGGLETON, C.V. KHEN and K.M. YUSAH. 2011. *Establishing the evidence base for maintaining biodiversity and ecosystem function in the oil palm landscapes of South East Asia.* Philosophical Transactions of the Royal Society B: Biological Sciences **366**:3277-3291.

GAVEAU, D.L., M. KSHATRIYA, D. SHEIL, S. SLOAN, E. MOLIDENA, A. WIJAYA, S. WICH, M. ANCRENAZ, M. HANSEN and M. BROICH. 2013. *Reconciling forest conservation and logging in Indonesian Borneo.* PLoS ONE **8**:e69887.

GAVEAU, D.L., A.S. SLOAN, E. MOLIDENA, HUSNAYANEM, M. ANCRENAZ, R. NASI, N. WIELAARD and E. MEIJAARD. 2014. *Four decades of forest persistence, loss and logging on Borneo.* PLoS ONE. **9**:e101654.

GIBSON, L., T.M. LEE, L.P. KOH, B.W. BROOK, T.A. GARDNER, J. BARLOW, C.A. PERES, C.J.A. BRADSHAW, W.F. LAURANCE, T.E. LOVEJOY and N.S. SODHI. 2011. *Primary forests are irreplaceable for sustaining tropical biodiversity.* Nature **478**:378-381.

GINGOLD, B., A. ROSENBARGER, Y.I.K.D. MULIASTRA, F. STOLE, I.M. SUDANA, M.D. MANESSA, M.A. MURDIMANTA, S. B. TIANGGA, C.C. MADUSARI and P. DOUARD. 2012. *How to identify degraded land for sustainable palm oil in Indonesia.* World Resources Institute and Sekala, Washington D.C.

GREGORY, S.D., B.W. BROOK, B. GOOSSENS, M. ANCRENAZ, R. ALFRED, L.N. AMBU and D.A. FORDHAM. 2012. *Long-term field data and climate-habitat models show that orangutan persistence depends on effective forest management and greenhouse gas mitigation.* PLoS ONE **7**:e43846.

HANSEN, M.C., P.V. POTAPOV, R. MOORE, M. HANCHER, S.A. TURUBANOVA, A. TYUKAVINA, D. THAU, S.V. STEHMAN, S. J. GOETZ, T.R. LOVELAND, A. KOMMAREDDY, A. EGOROV, L. CHINI, C.O. JUSTICE and J.R.G. TOWNSHEND. 2013. *High-resolution global maps of 21st-century forest cover change.* Science **342**:850-853.

HIJMANS, R.J., S.E. CAMERON, J.L. PARRA, P.G. JONES and A. JARVIS. 2005. *Very high resolution interpolated climate surfaces for global land areas.* International Journal of Climatology **25**:1965-1978.

HUSSON, S.J., S.A. WICH, A.J. MARSHALL, R.D. DENNIS, M. ANCRENAZ, R. BRASSEY, M. GUMAL, A.J. HEARN, E. MEIJAARD, T. SIMORANGKIR and I. SINGLETON. 2009. *Orang-utan distribution, density, abundance and impacts of disturbance.* Pages 77-96. S. Wich, S.U. Atmoko, T. Mitra Setia and C.P. van Schaik, editors. *Orang-utans: Geographic variation in behavioral ecology and conservation.* Oxford University Press, New York.

Institute for Development Studies (Sabah). 2007. *Sabah Development Corridor.* Sabah, Malaysia.

IPCC. 2013. *Climate change 2013: The physical science basis.* Working group 1 contribution to the fifth assessment report of the Intergovernmental Panel on Climate Change. Summary for Policymakers. IPCC, Switzerland.

IUCN. 2013. *IUCN Red list of threatened species. Version 2013.1.* World conservation Union (IUCN), Gland, Switzerland. www.iucnredlist.org. downloaded on 10 September 2013.

JETZ, W., D.S. WILCOVE and A.P. DOBSON. 2007. *Projected impacts of climate and land-use change on the global diversity of birds.* PLoS Biology **6**:1211-1219.

KOH, L.P. 2007. *Impending disaster or sliver of hope for Southeast Asian forests? The devil may lie in the details.* Biodiversity and Conservation **16**:3935-3938.

KOH, L.P., J. MIETTINEN, S.C. LIEW and J. GHAZOUL. 2011. *Remotely sensed evidence of tropical peatland conversion to oil palm.* Proceedings of the National Academy of Sciences **108**:5127-5132.

KOH, L.P. and N.S. SODHI. 2010. *Conserving Southeast Asia's imperiled biodiversity: scientific, management, and policy challenges.* Biodiversity and Conservation **19**:913-917.

LACKMAN, I. and A. MURPHY. 2014. *Legal mechanisms for orangutan habitat conservation outside of protected areas in Sabah, Malaysia.* Unpublished report by HUTAN.

LEADLEY, P. 2010. *Biodiversity scenarios: Projections of 21st century change in biodiversity, and associated ecosystem services: a technical report for the global biodiversity outlook 3.* UNEP/Earthprint.

MACKINNON, K., G. HATTA, H. HALIM and A. MANGALIK 1996. *The ecology of Kalimantan, Indonesian Borneo.* Periplus Editions, Singapore.

MANTEL, S., H. WÖSTEN and J. VERHAGEN. 2007. *Biophysical land suitabililty for oil palm in Kalimantan, Indonesia.*

ISRIC – World Soil Information, Alterra, Plant Research International, Wageningen UR, Wageningen, Netherlands.

MARSHALL, A.J., R. LACY, M. ANCRENAZ, O. BYERS, S. HUSSON, M. LEIGHTON, E. MEIJAARD, N. ROSEN, I. SINGLETON, S. STEPHENS, K. TRAYLOR-HOLZER, S.U. ATMOKO, C.P. VAN SCHAIK and S.A. WICH. 2009. *Orangutan population biology, life history, and conservation. Perspectives from population viability analysis models.* Pages 311-326 in S.A. WICH, S.U. ATMOKO, T. MITRA SETIA and C.P. VAN SCHAIK, editors. *Orang-utans: geographic variation in behavioral ecology and conservation.* Oxford University Press, Oxford, UK.

MEIJAARD, E., N.K. ABRAM, J.A. WELLS, A.-S. PELLIER, M. ANCRENAZ, D.L.A. GAVEAU, R.K. RUNTING and K. MENGERSEN. 2013. *People's perceptions about the importance of forests on Borneo.* PLoS ONE **8**:e73008.

MEIJAARD, E., D. BUCHORI, Y. HADIPRAKOSO, S.S. UTAMI-ATMOKO, A. TJIU, D. PRASETYO, NARDIYONO, L. CHRISTIE, M. ANCRENAZ, F. ABADI, I.N.G. ANTONI, D. ARMAYADI, A. DINATO, ELLA, P. GUMELAR, T.P. INDRAWAN, KUSSARITANO, C. MUNAJAT, A. NURCAHYO, C.W.P. PRIYONO, Y. PURWANTO, D.P. SARI, M.S.W. PUTRA, A. RAHMAT, H. RAMADANI, J. SAMMY, D. SISWANTO, M. SYAMSURI, J. WELLS, H. WU and K. MENGERSEN. 2011. *Quantifying killing of orangutans and human-orangutan conflict in Kalimantan, Indonesia.* PLoS ONE **6**:e27491.

MEIJAARD, E. and D. SHEIL. 2008. *The persistence and conservation of Borneo's mammals in lowland rain forests managed for timber: observations, overviews and opportunities.* Ecological Research **23**:21-34.

MEIJAARD, E. and D. SHEIL. 2013. *Oil-palm plantations in the context of biodiversity conservation.* Pages 600-612 in S.A. LEVIN, editor. Encyclopedia of Biodiversity. Academic Press, Waltham, MA.

MEIJAARD, E., A. WELSH, M. ANCRENAZ, S. WICH, V. NIJMAN and A.J. MARSHALL. 2010. *Declining orangutan encounter rates from Wallace to the present suggest the species was once more abundant.* PLoS ONE **5**:e12042.

MEIJAARD, E. and S. WICH. 2007. *Putting orang-utan population trends into perspective.* Current Biology **17**:R540.

MEIJAARD, E., S. WICH, M. ANCRENAZ and A.J. MARSHALL. 2012. *Not by science alone: why orangutan conservationists must think outside the box.* Annals of the New York Academy of Sciences **1249**:29-44.

MYERS, N., R.A. MITTERMEIER, C.G. MITTERMEIER, G.A.B. DA FONSECA and J. KENT. 2000. *Biodiversity hotspots for conservation priorities.* Nature **403**:853-858.

NELLEMANN, C., L. MILES, P.B. KALTENBORN, M. VIRTUE, and H. AHLENIUS. 2007. *The last stand of the orangutan. State of emergency: Illegal logging, fire and palm oil in Indonesia's national parks.* United Nations Environment Programme, GRID-Arendal: Arendal, Norway.

OLDEMAN, L. and I. LAS. 1980. *The agroclimatic maps of Kalimantan, Maluku, Irian Jaya and Bali, West and East Nusa Tenggara.* Contributions-Central Research Institute for Agriculture (Indonesia).

PARMESAN, C. 2006. *Ecological and evolutionary responses to recent climate change.* Annual Reviews of Ecology, Evolution and Systematics **37**:637-669.

RIJKSEN, H.D. and E. MEIJAARD. 1999. *Our vanishing relative. The status of wild orang-utans at the close of the twentieth century.* Kluwer Academic Publishers, Dordrecht, The Netherlands.

RUSSON, A.E. 2009. *Orang-utan rehabilitation and reintroduction: Successes, failures, and role in conservation.* Pages 327-350. *Orang-utans: Geographic variation in behavioral ecology and conservation.* Oxford University Press, New York.

Sabah Wildlife Department. 2011. *Orangutan Action Plan.* Kota Kinabalu, Sabah, Malaysia.

SALA, O.E., F.S. CHAPIN, III, J.J. ARMESTO, E. BERLOW, J. BLOOMFIELD, R. DIRZO, E. HUBER-SANWALD, L.F. HUENNEKE, R.B. JACKSON, A. KINZIG, R. LEEMANS, D.M. LODGE, H.A. MOONEY, M. OESTERHELD, N.L. POFF, M.T. SYKES, B.H. WALKER, M. WALKER and D.H. WALL. 2000. *Global biodiversity scenarios for the year 2100.* Science **287**:1770-1774.

SMIT, H.H., E. MEIJAARD, C. VAN DER LAAN, S. MANTEL, A. BUDIMAN and P. VERWEIJ. 2013. *Breaking the link between environmental degradation and oil palm expansion: A method for enabling sustainable oil palm expansion.* PLoS ONE **8**:e68610.

SODHI, N.S., L.P. KOH, B.W. BROOK and P.K.L. NG. 2004. *Southeast Asian biodiversity: an impending disaster.* Trends in Ecology & Evolution **19**:654-660.

STILES, D., I. REDMOND, D. CRESS, C. NELLEMANN, R.K. FORMO (eds). 2013. *Stolen apes – The illicit trade in chimpanzees, gorillas, bonobos and orangutans.* A Rapid Response Assessment. United Nations Environment Programme, GRID-Arendal.

STRUEBIG, M.J., M. FISCHER, D.L.A. GAVEAU, E. MEIJAARD, S.A. WICH, C. GONNER, R. SYKES, A. WILTING, and S. KRAMER-SCHADT. 2015. *Anticipated climate and land-cover changes reveal refuge areas for Borneo's orang-utans.* Global Change Biology. DOI: 10.1111/gcb.12814

SVENNING, J.-C. and R. CONDIT. 2008. *Biodiversity in a warmer world.* Science **322**:206-207.

TEWKSBURY, J.J., R.B. HUEY and C.A. DEUTSCH. 2008. *Putting the heat on tropical animals.* Science **320**:1296-1297.

THUILLER, W. 2007. *Biodiversity: Climate change and the ecologist.* Nature **448**:550-552.

VENTER, O., E. MEIJAARD, H.P. POSSINGHAM, R. DENNIS, D. SHEIL, S. WICH, L. HOVANI and K. WILSON. 2009. *Carbon payments as a safeguard for threatened tropical mammals.* Conservation Letters **2**:123-129.

WELLS, J., E. MEIJAARD, N.K. ABRAM and S. WICH. 2013. *Forests, floods, people and wildlife on Borneo.* A report to UNEP.

WICH, S., D.L.A. GAVEAU, N. ABRAM, M. ANCRENAZ, A. BACCINI, S. BREND, L.M. CURRAN, R.A. DELGADO, A. ERMAN, G. FREDRIKSSON, B. GOOSSENS, S. HUSSON, I. LACKMAN, A.J. MARSHALL, A. NAOMI, E. MOLIDENA, NARDIYONO, A. NURCAHYO, K. ODOM, A. PANDA, PURNOMO, A. RAFIASTANTO, D. RATNASARI, A.H. SANTANA, I. SAPARI, C.P. VAN SCHAIK, J. SIHITE, S. SPEHAR, E. SANTOSA, A. SUYOKO, A. TIJU, G. USHER, S.S.U. ATMOKO, E.P. WILLEMS and E. MEIJAARD. 2012. *Understanding the impacts of land-use policies on a threatened species: Is there a future for the Bornean orang-utan?* PLoS ONE **7**:e49142.

WICH, S., RISWAN, J. JENSON, J. REFISCH and C. NELLEMAN. 2011. *Orangutans and the economics of sustainable forest management in Sumatra.* Page 83. UNEP/GRASP/PanEco/YEL/ICRAF/GRID-Arendal. Birkeland Trykkeri AS, Norway.

WICH, S.A., E. MEIJAARD, A.J. MARSHALL, S. HUSSON, M. ANCRENAZ, R.C. LACY, C.P. VAN SCHAIK, J. SUGARDJITO, T. SIMORANGKIR, K. TRAYLOR-HOLZER, M. DOUGHTY, J. SUPRIATNA, R. DENNIS, M. GUMAL, C.D. KNOTT and I. SINGLETON. 2008. *Distribution and conservation status of the orang-utan (Pongo spp.) on Borneo and Sumatra: How many remain?* Oryx:329-339.

WILSON, K.A., E. MEIJAARD, S. DRUMMOND, H.S. GRANTHAM, L. BOITANI, G. CATULLO, L. CHRISTIE, R. DENNIS, I. DUTTON and A. FALCUCCI. 2010. *Conserving biodiversity in production landscapes.* Ecological Applications **20**:1721-1732.